外教社 英汉·汉英百科词汇手册系列

总主编 庄智象

英汉·汉英
地球科学词汇手册

An English-Chinese Chinese-English
Glossary of Earth Science

编者 孙东云

上海外语教育出版社
SHANGHAI FOREIGN LANGUAGE EDUCATION PRESS

图书在版编目(CIP)数据

英汉·汉英地球科学词汇手册/孙东云编.—上海：
上海外语教育出版社，2012
(外教社英汉·汉英百科词汇手册系列)
ISBN 978-7-5446-2880-8

Ⅰ.①英… Ⅱ.①孙… Ⅲ.①地球科学－词汇－手册－英、汉 Ⅳ.①P-62

中国版本图书馆 CIP 数据核字(2012)第 194196 号

出版发行：	**上海外语教育出版社**
	（上海外国语大学内）　邮编：200083
电　　话：	021-65425300（总机）
电子邮箱：	bookinfo@sflep.com.cn
网　　址：	http://www.sflep.com.cn　http://www.sflep.com
责任编辑：	陈飘平
印　　刷：	上海华业印刷厂有限公司
开　　本：	787×965　1/32　印张 6.375　字数 275 千字
版　　次：	2012 年 11 月第 1 版　2012 年 11 月第 1 次印刷
印　　数：	3 000 册
书　　号：	ISBN 978-7-5446-2880-8 / P·0003
定　　价：	15.00 元

本版图书如有印装质量问题，可向本社调换

目 录

序 v

前言 viii

使用说明 ix

英汉部分 1～96

汉英部分 97～190

参考文献 191

外教社英汉·汉英百科词汇手册系列

总主编:庄智象

英汉·汉英地球科学词汇手册

编　　者:孙东云
策划编辑:张春明　李法敏
责任编辑:陈飘平

序

改革开放 30 年来,我国的经济、科技、文化、教育、外贸和社会各项事业快速发展,取得了令人瞩目的成就,经济总量翻了两番之多,国内生产总值 2007 年已近 25 万亿元人民币,人均国民收入达到了 2 000 多美元,提前实现了小康目标;科技水平不断提高,高新技术快速发展,缩短了与世界先进国家的差距;文化、教育事业迅猛发展,不断满足和丰富着人民日益增长的文化精神需要;社会服务和保障体系不断完善,使中华民族和社会更加和谐。我国在国际事务中的作用日益凸显,与世界各国的政治、经济、科技、文化、教育、外交、军事等方面的交往日益频繁。成功加入世界贸易组织,成功申办和举办 2008 年北京奥运会和残奥会,成功申办上海世博会等一系列重大外交活动,塑造了中国在世界上的良好形象,更是促进了我国与世界各国的交往、交流和交融。经济全球化、科技一体化、文化多元化、信息网络化的发展趋势,使人们的生活正发生着深刻的变化。

新的学科、新的术语、新的词汇不断诞生和涌现。我国数亿不同层次的英语学习者和使用者,在学习或工作中不断遇到新的词语、新的术语。大部分的英语学习者和使用者能够比较顺利地用普通英语进行交流和交际,而一旦遇到专业领域的词汇或术语往往会陷入困境,有时可能了解某词语的一般意义,但不知道它在某些专业领域指的是什么。这种情况,在日常交往中,或在专业性较强的国际会议中屡见不鲜,常常需要英语专家和专业方面的人士通力合作,才能解疑释惑。若能编纂出版一套英汉、汉英的百科词汇手册,则将大大有益于英语学习者和使用者,能够为他们的英语学习和使用助上一臂之力。笔者曾在上世纪90年代初随同中国教材出版项目考察团访美,任翻译,就是依靠了一本英汉汉英出版印刷词汇手册,比较顺利地完成了任务。同时,此类专业词汇手册还有助于学习者提高英语水平和能力,借助英语汲取科技知识和信息,扩大视野,不断丰富和提高专业知识和水平。有鉴于此,上海外语教育出版社(以下简称外教社)与牛津大学出版社合作,于本世纪初陆续出版了牛津百科分类词典英语版40余种。这批百科词典的出版深受专业人员、英语学习者和使用者的欢迎。同时,根据部分读者的要求,外教社经过充分调研和论证,并征得牛津大学出版社同意,从该系列词典中挑选出近十种,组织有关专业人员编译成英汉双解版;由于工作量巨大,亦仅将这部分词典的词目翻译成汉语。随着对外交流和交往的深入发展,不断有读者询问外教社是否出版有完整的百科方面的词汇手册或词典。为满足这一需要,外教社经过广泛的调研、需求分析和市场论证,组织编写了外教社英汉·汉英百科词汇手册系列图书,历经四年的努力,全国近百位编纂者的通力合作和辛勤劳动,终于迎来了第一批书稿。

本百科词汇手册系列有以下几个特点:

1. 按学科、专业和行业分册编写(以下统称专业),分类依据主要是国家标准学科分类、国家标准国民经济行业分类、企业经营行业分类及中国图书在版编目分

类,涵盖自然科学、技术、社会科学、人文科学、社会生活等80余个领域;

2. 各专业词汇手册包括英汉、汉英两部分,尽可能涵盖各专业最基本、最常用的词汇,每部分收词基本上控制在5 000至10 000条,版式和开本设计便于使用和携带;

3. 各分册按词汇的使用频率收列专业基本词汇,同时力求反映该专业的最新发展,只收专业性词汇和普通语文词汇的专业性义项。与本专业关系不大的其他专业词汇尽量不收。

本系列词汇手册可供英语学习者、使用者及相关专业人员了解和熟悉专业词汇,学习和丰富专业知识,提升专业视野和水平之参考,亦可作为翻译的参考工具书。由于学识和能力所限,新的词汇层出不穷,收词难免挂一漏万,谬误和缺陷在所难免,敬请广大读者惠予批评、指正。

庄智象
2008年10月

前 言

地球科学是以地球系统的过程与变化及其相互作用为研究对象的基础学科,其分支学科包括地理学、地质学、地球物理学、地球化学、大气科学、海洋科学、空间物理学、地球信息科学等,与很多自然学科都有关联,而且在防灾减灾、空间探索、资源勘探等领域都发挥着重要作用。近30年来,地球科学发展日新月异,大量新术语不断涌现,另一方面,国际学术交流活动也日渐频繁,这就对地球科学研究者提出了语言方面的要求,也迫切需要一本能够反映地球科学最新发展的双语工具书。但目前图书市场上该类工具书却很少,鉴于此,我们编写了这本《英汉·汉英地球科学词汇手册》。

本词汇手册英汉、汉英双向收词11 000余条,内容涵盖了地球科学以及相关学科如地理学、地质学、大气科学、海洋科学等学科的基础词汇,既收录单词,也收录以单词为基础的多词术语。在术语的翻译上,笔者以全国自然科学名词审定委员会公布的术语表和相关国家标准为基础,并参阅了大量其他相关词汇手册,力争做到有章可循,保证其权威性和可靠性,并兼顾行业内习惯译法。近年来,地球科学的发展迅猛,新术语不断涌现,有很多尚未收录到现有资料当中,因此笔者查阅了各种相关资料,采用业界最广为接受的译法。

地球科学上至天文,下至地理,而新世纪科研发展的跨学科性使得编纂这本术语手册变得异常困难。由于笔者水平有限,难免在词条分布、译名选择等方面存在一定的错误和疏漏,还请读者和大方之家多多批评指正。

使 用 说 明

1. 本手册分为英汉、汉英两部分。
2. 英汉部分条目按照字母顺序排列;汉英部分条目按照拼音音序排列,同音字按照笔画数排列。
3. 英汉部分以数字、特殊符号、希腊字母等非英语字母开头的条目放在英汉部分正文之后;汉英部分以数字、特殊符号、西文字母等非汉字开头的条目放在汉英部分正文之后。
4. 外国人名条目的词头与译文一般采用姓前名后的形式。例如:
    Chomsky, Avram Noam 乔姆斯基,A·N
    乔姆斯基,A·N Chomsky, Avram Noam
5. 为便于查阅,汉英部分的某些条目(多为自由组合)以汉语中心词或关键词立目,其他文字置于空心鱼尾号中,放在中心词或关键词之后。例如:
    与产品相关的行为变量　product-related behavioural descriptor
    立目时处理为:
    行为变量〖与产品相关的〗　product-related behavioural descriptor
6. 英汉部分概念相同或相近的汉语译文之间用逗号分隔,概念不同的汉语译文之间用分号分隔;汉英部分的英语译文一律用分号间隔。
7. 圆括号"( )"用于括注:① 缩写或全称形式;② 可省略的内容;③ 解释说明性文字。
8. 方括号"[ ]"用于括注可替换的文字。
9. 尖括号"〈 〉"用于括注学科标注。

英汉部分

Abbe comparator principle 阿贝比长原理
aberration 像差
aberration constant 像差常数
ablation 烧蚀
ablation of meteorites 陨星烧蚀
abscissa 横坐标
absolute black body 绝对黑体
absolute brightness 绝对亮度
absolute calibration 绝对校准
absolute error 绝对误差
absolute flying height 绝对航高
absolute gravimeter 绝对重力仪
absolute gravity measurement 绝对重力测量
absolute intensity 绝对强度
absolute magnitude 绝对星等
absolute orientation 绝对定向
absolute perturbation 绝对摄动
absolute scintillation efficiency 绝对闪烁效率
absolute solar flux 绝对太阳通量
absolute temperature 绝对温度
absolute threshold 绝对阈
absolute time 绝对时间
absolute value 绝对值
absolute vorticity 绝对涡度
absolute zero 绝对零度
absorbing medium 吸收介质
absorption band 吸收带
absorption coefficient 吸收系数
absorption index 吸收指数
absorption spectrum 吸收光谱
abstract symbol 抽象符号
abundance of elements 元素丰度
abundance value 丰度值
accelerating mechanism 加速机制
acceleration potential 加速度势
acceleration response spectrum 加速度反应谱
accelerograph 加速度仪
accelerometer 加速度计
acceptance cone 接纳锥
accidental count 偶然计数
accidental error 偶然误差
AC demagnetization (alternating current demagnetization) 交流退磁
acid geochemical barrier 酸性地球化学障
aclinic line 零倾线,无倾线
acoustic frequency 声频
acoustic-gravity wave 声重力波
acoustic logging 声波测井
acoustic velocity 声速
acoustic wave 声波
acquisition 捕获

acquisition and tracking radar 捕获和跟踪雷达
actinic absorption 光化吸收
actinograph 辐射仪,日光强度自动记录器
activated band 活动光带
activation energy 活化能
active aurora 活动极光
active fault 活动断层
active guidance 主动制导
active phase 活跃期
active prominence region 活动日珥区
active source method 主动源(方)法
active sun phenomena 活动太阳现象
active sunspot prominence 活动黑子日珥
actual field intensity 实际场强
adaptation level 适应性水平
adaptive stack 自适应叠加
additional potential 附加位
addition constant 加常数
adiabatic acceleration 绝热加速
adiabatic adjustment 绝热调节
adiabatic redistribution 绝热再分布
adit planimetric 坑道平面图
adit prospecting engineering survey 坑探工程测量
adjusted value 平差值
adjustment 平差;调节
adjustment by method of junction point 结点平差
adjustment calculation 平差计算
adjustment method 平差法
adjustment model 平差模型
adjustment of correlated observation 相关(观测)平差
adjustment of observation 测量平差
administrative map 行政区划图
adopted latitude 纬度采用值
adopted longitude 经度采用值
AE index (auroral electrojet index) 极光带电集流指数,AE 指数
AEM method (airborne electromagnetic method) 航空电磁法
AEM system (airborne electromagnetic system) 航空电磁系统
aeration 通风;掺气
aerial 航空的
aerial camera 航空摄影机
aerial film 航摄软片
aerial gravity measurement 航空重力测量
aerial image 航空影像
aerial photograph 航摄像片
aerial photographic gap 航摄漏洞
aerial remote sensing 航空遥感
aerial survey 航测
aeroballistics 航空弹道学
aerobiology 高空生物学,航空生物学
aeroclimatology 高空气候学
aerodynamic 空气动力的
aerogeochemical prospecting 航空地球化学勘探
aerogram 高空图解
aeroleveling 空中水准测量
aerology 高空气象学
aeromagnetic 航空磁测的
aeromagnetic survey 航空磁测
aeronautical chart 航空图
aeronomy 高空大气学
aerophotogonometry 空中导线测量
aerosol 悬浮颗粒
aerotriangulation 空中三角测量
AF cleaning (alternating field cleaning) 交变场清洗
affine 仿射的
affine plotting 仿射绘图,变换光束测图
AFMAG (audio frequency magnetic field method) 音频磁场法

after-effect of (magnetic) storm 磁暴后效
afterglow 余辉
aftershock 余震
age of remanence 剩磁年龄
aggregate 聚集体
agonic line 零偏线
agravic 失重的
agricultural geochemistry 农业地球化学
air afterglow emission 大气余辉发射
airborne 航空的;机载的
airborne electromagnetic method (AEM method) 航空电磁法
airborne electromagnetic system (AEM system) 航空电磁系统
airborne geochemical prospecting 航空地球化学勘探
airborne gravimeter 航空重力仪
airborne gravity measurement 航空重力测量
airborne laser sounding 机载激光测深;机载激光测探
airborne magnetic survey 航空磁测
airborne radioactivity survey 航空放射性测量
airborne sensor 机载遥感器
air-coupled Rayleigh wave 空气耦合瑞利波
airfield runway survey 机场跑道测量
Air Force atmospheric model (美国)空军大气模式
airglow 气辉
airglow brightness 气辉亮度
airglow emission 气辉发射
air gun 空气枪(震源)
air launch 空中发射
air pollution 空气污染,大气污染
airport survey 机场测量
air shower 大气簇射
air wave 空气波
Airy-Heiskanen isostasy 艾里-海斯卡宁均衡

Airy phase 艾里震相
AKR (auroral kilometric radiation) 极光千米波辐射
albedo 反照率
albedo electron 反照电子
albedo neutron 反照中子
albedo neutron spectrum 反照中子能谱
albedo neutron theory 反照中子理论
albedo of the Earth 地球反照率
Alfvén layer 阿尔文层
Alfvén Mach number 阿尔文马赫数
Alfvén perturbation theory 阿尔文微扰理论
algorithm 算法
alidade 视准仪;测高仪
alkali metal 碱金属
alkali metal atom 碱金属原子
alkaline geochemical barrier 碱性地球化学障
alkaline hydrolysis 碱解
all-inertial guidance 全惯性制导
all-sky camera 全天空照相机
all-sky photometer 全天空光度计
alluvial 冲积的
almanac (天文)年历
almucantar 地平纬圈,等高圈
alpha energy loss α能量损失
alpha particle α粒子
alpha ray α射线
altazimuth 地平经纬仪
alternate maxima 交替极大值
alternate minima 交替极小值
alternating current demagnetization (AC demagnetization) 交流退磁
alternating field cleaning (AF cleaning) 交变场清洗
altimeter 测高仪
altimetry 测高(法)
altitude angle 高度角
altitude correction of zenith difference 异顶差

altostratus cloud 高层云
Amalthea 木卫五
ambiguity 模糊度
ambiguity resolution 模糊度解算
ambipolar diffusion 双极扩散
amino acid geochemistry method 氨基酸地球化学法
ammonia 氨
ammonia ionization limit 氨电离限
amphibious seismic operation 两栖地震勘探作业
amplitude 振幅
amplitude envelope 振幅包络
amplitude scintillation 幅度闪烁
anaglyphical stereoscopic viewing 互补色立体观察
anaglyphic map 互补色地图
anaglyphoscope 互补色镜
analog aerotriangulation 模拟空中三角测量
analog magnetometer 模拟磁强计
analog map 模拟地图
analog photogrammetric plotting 模拟法测图
analog photogrammetry 模拟摄影测量
analog stereoplotter 模拟立体测图仪
analog tape 模拟磁带
analysis of structural response to seismic excitation 地震响应分析
analytical aerotriangulation 解析空中三角测量
analytical geochemistry 分析地球化学
analytical map 分析地图
analytical mapping 解析测图
analytical orientation 解析定向
analytical photogrammetry 解析摄影测量
analytical plotter 解析测图仪
analytical rectification 解析纠正
analytical solution of motion equation 运动方程分析解

analytic mapping control point 解析图根点
anaseism 离源震
anaseismic onset 离源初动
anchorage 锚地
anchorage berth 锚位
anchorage-prohibited area 禁锚区
ancient geothermal system 古地热系统
ancient map 古地图
aneroid barometer 空盒气压表
aneroidograph 空盒气压计
angle modulation 调角
angle of arrival 到达角;落角
angle of attack 攻角;冲角;迎角
angle of declination 偏角
angle of depression 俯角
angle of deviation 偏向角
angle of elevation 仰角
angle of inclination 倾角
angle of lag 落后角,滞后角
angle of minimum deviation 最小偏差角,最小偏向角
angle of roll 侧滚角
angle of spread 展开角
angle of torsion 扭转角
angular acceleration 角加速度
angular field of view 像场角
anhysteretic remanent magnetization (ARM) 无滞剩磁
animated mapping 动画制图
animated steering 动画引导
anisotropic 各向异性的
anisotropic conductivity 各向异性传导率;各向异性电导率
anisotropic medium 各向异性电介质
anisotropic scattering 各向异性散射
annihilation 湮没
annihilation of magnetic field energy 磁(场)能(量)湮没
Anno Domini 公元
annual change of magnetic variation 磁变年差

annual mean 年均值
annual mean sea level 年平均海面
annular eclipse （日）环食
anode 阳极
anomalous dispersion 反常色散
anomalous geochemical gradient 异常地球化学梯度
anomalous ionization 异常电离
anomalous tail 反常彗尾
anomaly 异常
Antarctic ozone distribution 南极臭氧分布
antenna 天线
antenna aperture 天线孔径
antenna impedance 天线阻抗
antenna matching 天线匹配
anthracene 蒽
anticentre 震中对跖点
anticlockwise 逆时针的
anticoincidence circuit 反符合电路
anticorrelation 负相关
anticyclone 反气旋
anticyclonic ridge 反气旋脊；高压脊
anti-epicentre 震中对跖点
antigravity 抗重力；无重量
anti-plane shear crack 反平面剪切裂纹
antiroot 反山根
anti-seismic design of main power buildings 主厂房抗震设计
antisymmetrical mode 反对称振型
aperture 孔，孔径
aperture angle 孔径角
aperture ratio 孔径比，口径比
aphelion 远日点
Ap index Ap 指数
apoapsis 远（拱）点
apocynthion 远月点
apogee 远地点
apparent height 视高
apparent magnetic susceptibility 视磁化率

apparent magnitude 视星等
apparent polar wander 视极移
apparent polar-wander curve 视极移曲线
apparent polar-wander path (APWP) 视极移路径
apparent resistivity 视电阻率
apparent stress 视应力
apparent velocity 视速度
Appleton anomaly 阿普尔顿异常
Appleton-Hartree formula 阿普尔顿-哈特里公式
Appleton layer 阿普尔顿层
applied cartography 实用地图学
applied frequency 应用频率
applied geochemistry 应用地球化学
applied geophysics 应用地球物理(学)
applied seismology 应用地震学
approximate adjustment 近似平差
appulse 半影月食
apsidal angle 拱心角
apsidal motion 拱线运动
APWP (apparent polar-wander path) 视极移路径
aquifer 含水层
arbitrary projection 任意投影
arc discharge 电弧放电，弧光放电
archaeological photogrammetry 考古摄影测量
archaeomagnetism 考古地磁(学)
Archimedes spiral 阿基米得螺线
architectural photogrammetry 建筑摄影测量
arc measurement 弧度测量
Arctic 北极
Arctic Circle 北极圈
arc-to-chord correction in Gauss projection 高斯投影方向改正
area coverage 覆盖区，影响区
area leveling 面水准测量
area method 范围法
area symbol 面状符号

areocentric 火星中心的
areography 火星表面学,火面学
areophysics 火星物理学
argon 氩
argument of perigee 近地点角距
arid area 干旱区
Ariel 天卫一
ARM (anhysteretic remanent magnetization) 无滞剩磁
array 台阵
array factor 排列系数
arrival time 到时
arrival time difference 到时差
arrowhead method 运动线法
arrow plot 箭头图
artificial airglow 人造气辉
artificial aurora 人造极光
artificial comet 人造彗星
artificial earthquake 人工地震
artificial emission 人工发射
artificial magnetization method 人工磁化法
artificial satellite 人造卫星
artificial seismic source 人工震源
ascending node 升交点
ascending pass 升轨
aseismic belt 无震带
aseismic joint 防震缝
aseismic ridge 无震海岭
aseismic slip 无震滑动
aseismic zone 无震区
ashen light 金星灰光
asiderite 石陨星,无铁陨石
asperity 凹凸体
asperity source model 凹凸体震源模式
associative detachment 结合性分离,缔合脱离
associative perception 整体感
assumed coordinate system 假定坐标系
assumed latitude 选择纬度,假定纬度
assumed longitude 选择经度,假定经度

astatic gravimeter 助动重力仪
astatic magnetometer 无定向磁强计
asteroid 小行星
asteroid belt 小行星带
asteroid zone 小行星带
asthenosphere 软流层
astroballistics 天体弹道学
astrodynamics 天体动力学
astro-geodetic deflection of the vertical 天文大地垂线偏差
astro-geodetic network 天文大地网
astronautics 航天学,星际航行学
astronavigation 天文导航
astronomical 天文的
astronomical constant 天文常数
astronomical coordinate 天文坐标
astronomical latitude 黄纬,天文纬度
astronomical longitude 黄经,天文经度
astronomical meridian 天文子午圈;天文子午线
astronomical point 天文点
astronomical refraction 天文折射
astronomical scintillation 天文闪烁
astronomical spectrograph 天体摄谱仪
astronomical spectroscopy 天体光谱学
astronomical time 天文时
astronomical unit 天文单位
astrophysics 天体物理学
astrospectroscopy 天体光谱学
asymmetric effect 东西效应;非对称效应
asymptotic direction 渐近方向
asymptotic latitude 渐近纬度
asymptotic longitude 渐近经度
Atlas 土卫十五
atlas 地图集
atlas information system 地图集信息系统

atmosphere optical thickness 大气光学厚度
atmosphere scale height 大气标高
atmosphere zenith delay 大地天顶延迟
atmospheric absorption 大气吸收
atmospheric band 大气谱带
atmospheric boundary 大气边界
atmospheric braking 大气制动
atmospheric circulation 大气环流
atmospheric density 大气密度
atmospheric diffusion 大气扩散
atmospheric electricity 大气电学
atmospheric emission 大气发射
atmospheric extinction 大气消光
atmospheric ion 大气离子
atmospheric ionization 大气电离
atmospheric model 大气模式
atmospheric opacity 大气不透明度,大气浑浊度
atmospheric oscillation 大气振荡
atmospheric parameter 大气参数
atmospheric pressure 大气压
atmospheric process 大气过程
atmospheric radiation 大气辐射
atmospheric refraction 大气折射
atmospheric response 大气响应
atmospheric structure 大气结构
atmospheric tide 大气潮汐
atmospheric turbidity 大气浑浊度
atmospheric turbulence 大气湍流
atmospheric vorticity 大气涡度
atmospheric window 大气窗
atmospheric window region 大气窗区
atomic absorption coefficient 原子吸收系数
atomic clock 原子钟
atomic emission 原子发射
atomic hydrogen 原子氢
atomic hydrogen chemistry 原子氢化学
atomic mass number 原子质量数
atomic oxygen 原子氧
atomic precession magnetometer 原子进动磁强计
atomic rocket 原子火箭
atom reaction 原子反应
attached shock wave 附着激波,附体激波
attachment coefficient 附着系数
attenuation 衰减
attenuation coefficient 衰减系数
attenuation constant 衰减常数
attitude 姿态
attitude control 姿态控制,姿控
attitude gyro 姿态陀螺仪
attitude of a satellite 卫星姿态
attribute 属性
attribute accuracy 属性精度
attribute testing 属性检定
audio frequency magnetic field method (AFMAG) 音频磁场法
Auger shower 俄歇簇射,奥格簇射
aurora 极光
aurora activity 极光活动
aurora australis 南极光
aurora borealis 北极光
auroral arc 极光弧
auroral belt 极光带
auroral cloud 极光云
auroral corona 极光冕
auroral echo 极光回波
auroral electrojet 极光带电集流
auroral electrojet index (AE index) 极光带电集流指数,AE指数
auroral electron 极光电子
auroral kilometric radiation (AKR) 极光千米波辐射
auroral line 极光谱线
auroral luminescence 极光发光
auroral morphology 极光形态学
auroral oval 极光卵形环
auroral particle 极光粒子
auroral pattern 极光形式
auroral physics 极光物理学
auroral radar 极光雷达
auroral spectrum 极光光谱

aurora polaris 北极光
aurora proton 极光质子
Authoritative Topographic Cartographic Information System （德国）官方地形制图信息系统
autocorrelation 自相关
autocorrelation coefficient 自相关系数
auto-covariance 自协方差
autokinetic effect 动感
automatic cartography 自动化地图制图
automatic celestial navigation 自动天文导航
automatic generalization 自动综合
automatic multifrequency ionospheric recorder 自动多频电离层记录仪
automatic plotting 自动绘图
automatic standard magnetic observatory 自动标准地磁观测台

autumnal equinox 秋分（点）
average error 平均误差
average lag 平均滞后
average response 平均响应
average solar wind speed 平均太阳风速
aviation astronomy 航空天文学
Avogadro number 阿佛伽德罗数
away polarity 背阳极性
away sector 背阳扇区
axis pole 轴极
azimuth 方位，方位角，地平经度
azimuthal component 方位角分量
azimuthal drift 方位漂移
azimuthal projection 方位投影
azimuthal symmetry 方位对称性
azimuth circle 方位圈，地平圈
azimuth distance positioning method 极坐标定位方法，方位距离定位方法
azimuth of photograph 像片方位角

Bb

baked contact test 烘烤接触检验
ballistics 弹道学
ball lightning 球状闪电
balloon-borne detector 球载探测器
balloon-borne sonde 球载探针
balloon observation 气球观测
balloon satellite 气球卫星
balloon-sonde 探测气球
Balmer discontinuity 巴耳末跳跃,巴耳末间断
Balmer formula 巴耳末公式
bandspread 频带展宽
band width 带宽
bank seismic facies 滩状地震相
bare rock 明礁
barium cloud 钡云
barium vapour 钡蒸汽
barn 靶
baroclinic 斜压的
baroclinic condition 斜压条件
barogram 气压图
barograph 气压计,气压记录器
barometer 气压计
baropause 气压层顶
barosphere 气压层
barotropic condition 正压情况
barotropic fluid 正压流体
barotropic state 正压状态
barotropy 正压(性)
barrier 障碍体
barrier source model 障碍体震源模式
barycentre 引力中心,质心
barye 微巴
base-height ratio 基-高比
baseline 基线
baseline flying 基线飞行
baseline measurement 基线测量
baseline network 基线网
base station 岸台,固定台
basic geochemical law 地球化学基本定律
basic geochemical map 基本地球化学图
basic gravimetric point 基本重力点
basic scale 基本比例尺
basin 盆地;流域
B-axis B轴;N轴;零向量
bay cable 浅海海底电缆
Bayesian classification 贝叶斯分类
beacon 信标;立标
beacon delay 信标延迟
beacon satellite 信标卫星
beam angle 波束角
beam antenna 波束天线,定向天线
beam efficiency 波束效率,射束效率
beam riding 束导
beam steering 延时组合
beat frequency 拍频,差频
bedding correction 层面改正
bedrock 基岩
Beer-Lambert law 比尔-兰伯特定律
Belinda 天卫十四
belt of totality 全食带
belt topographic map 带状地形图
benchmark 水准点
bend 弯道

bending method 弯曲法
Benioff seismograph 贝尼奥夫地震仪
Benioff zone 贝尼奥夫带
Bernoulli's theorem 伯努利定理
berth 泊位
Bessel ellipsoid 贝塞尔椭球
Bessel equation 贝塞尔方程
Bessel formula for solution of geodetic problem 贝塞尔大地主题解算公式
Bessel function 贝塞尔函数
beta decay β衰变
beta factor β因子
beta particle β粒子
beta ray β射线
betatron 电子感应加速器,电子回旋加速器
betatron effect 电子感应加速效应
between-sites precision 采点间精度
Bianca 天卫八
bidirectional anisotropy 双向异性
bidirectional flux 双向通量
bidirectionality 双向性
bilateral faulting 双侧断裂
bilateral geochemical barrier 两侧地球化学障碍
bi-Maxwellian distribution 双麦克斯韦分布
bimodal diffusion 双式扩散
binary code 二进制码
binary system 双重星系
binding 结合,键联
binding energy 结合能
bioastronautics 生物航天学,生物宇宙航行学
biogeochemical anomaly 生物地球化学异常
biogeochemical barrier 生物地球化学障
biogeochemical cycle of pollutants 污染物生物地球化学循环
biogeochemical disease 生物地球化学性疾病

biogeochemical prospecting 生物地球化学勘探
biogeochemistry 生物地球化学
biopak 生物容器,生物舱
biosphere 生物圈
bipolar group 双极群
bipolar magnetic region 双极磁区
bird's eye view map 鸟瞰图
Birkeland current 伯克兰电流
Bjerhammar problem 布耶哈马问题
black body 黑体
black hole 黑洞
blackout （短波通讯）中断
black soil 黑土
Blake event 布莱克事件
Blake excursion 布莱克漂移
BL coordinate BL 坐标
blind zone 盲区
block 块体
block adjustment 区域网平差
block diagram 块状图
blocking circulation 阻塞环流
blocking diameter 阻挡直径
blocking temperature 阻挡温度
blocking time 阻挡时间
blocking volume 阻挡体积
blue key 蓝底图
blue shift 蓝移
bodily seismic wave 地震体波
body wave magnitude 体波震级
Bohm criterion 博姆判据
boiling mud pool 沸泥塘
boiling spring 沸泉
bolide 火球,火流星
Boltzmann constant 玻耳兹曼常数
Boltzmann formula 玻耳兹曼公式
Bond albedo 邦德反照率
borehole 钻孔,井眼
borehole-borehole variant 井中-井中方式
borehole deformation gauge 钻孔形变计

borehole photo 井中摄影
borehole strainmeter 钻孔应变计
borehole stressmeter 钻孔应力计
borehole-surface variant 井中-地面方式
borehole televiewer 井中电视
Bosch-Omori seismograph 玻什-大森地震仪
bottom-side ionogram 底部频高图
bottom-side ionosphere 底部电离层
bottom-side sounder 底视探测仪
Bouguer anomaly 布格异常
Bouguer correction 布格改正
Bouguer reduction 布格校正
boundary condition 边界条件
boundary mark 界址点
boundary point 界址点
boundary velocity 界面速度
boundary wave 界面波
bound charge 束缚电荷
bound-free transition 束缚-自由跃迁
bow shock 弓激波
box classification method 盒式分类法
Boyle's law 玻意耳定律
braking orbit 制动轨道
branching ratio 分支比
breadboard 实验线路板
breakdown 击穿;崩溃;分类
breakout phase 突发相

Briden index 布利登指数
bridge axis location 桥梁轴线测设
bridge construction control survey 桥梁控制测量
bridge survey 桥梁测量
bridging of model 模型连接
bright flocculus 亮谱斑
brightness coefficient 亮度系数
brightness distribution 亮度分布
bright spot （地震勘探）亮点
Brno excursion 布尔诺漂移
broad spectral band 宽谱带
Browne correction 布朗热改正
brown soil 棕钙土棕壤
Brunhes epoch 布容期
Bruns formula 布隆斯公式
buffer 缓冲区
building aseismicity 建筑抗震
building axis survey 建筑物沉降观测
building engineering survey 建筑工程测量
bulge effect 隆起效应
bulk motion 整体运动
buoyancy frequency 浮力频率
Burg deconvolution 伯格反褶积
burnout altitude 燃尽高度,熄火高度
burst source 爆发源
butterfly diagram 蝶形图

C9 index C9 指数
CAC (computer-aided cartography) 机助地图制图
C/A Code (Coarse/Acquisition Code) 粗码
cadastral mapping 地籍制图
cadastral survey 地籍测量
cadastration 地籍测量
cadastre 地籍
cadographic analysis 地图分析
cadographic classification 地图分类
cadographic communication 地图传输
cadographic information 地图信息
Cagniard-De Hoop method [technique] 卡尼亚尔-德胡普法
Cagniard method 卡尼亚尔法
calcic geochemical barrier 钙地球化学障
calcium plage 钙谱斑
calibration 刻度;校准,标定
calibration event 主导事件
calibration seismic event 主导地震事件
calibration sphere 标定球
caliper survey 井径测井
Callisto 木卫四
Calypso 土卫十四
Canonical coordinate 正则坐标
cap prominence 冠状日珥
capture cross-section 捕捉截面
carbonate 碳酸盐
carbonated spring 碳酸泉
carbonate reservoir 碳酸盐岩类储集层
carbon cycle 碳循环
carbon dioxide 二氧化碳
carbon monoxide reaction 一氧化碳反应
carrier 载波
carrier frequency 载频
carrier phase measurement 载波相位测量
Cartesian coordinate 笛卡儿坐标
cartographic analysis 地图分析
cartographic classification 地图分类
cartographic communication 地图传输
cartographic evaluation 地图评价
cartographic exaggeration 制图夸大
cartographic expert system 制图专家系统
cartographic generalization 制图综合
cartographic hierarchy 制图分级
cartographic information 地图信息
cartographic information system (CIS) 地图信息系统
cartographic language 地图语言
cartographic methodology 地图研究法
cartographic model 地图模型
cartographic organization 地图内容结构
cartographic potential information 地图潜信息
cartographic pragmatics 地图语用

cartographic presentation 地图表示法
cartographic selection 制图选取
cartographic semantics 地图语义
cartographic semiology 地图符号学
cartographic simplification 制图简化
cartographic syntactics 地图语法
cartography 地图学，地图制图
cartometry 地图量算法
cartwheel satellite 滚轮式卫星
cascade migration 级联偏移
cascade shower 级联簇射
casing 套管
catchment area survey 汇水面积测量
cathode 阴极
cause of earthquake 地震成因
cave 洞穴
cavitation 空化
cavity resonance 空腔共振
CCT (computer compatible tape) 计算机兼容磁带
CDP grid (common-depth-point grid) 共深度点网格
CDP stacking (common-depth-point stacking) 共深度点叠加
celestial background 天体背景
celestial coordinate 天体坐标
celestial latitude 黄纬，天球纬度
celestial longitude 黄经，天球经度
celestial position 天体位置
cell 格网单元
cellular convection 环形对流
Cenozoic 新生代
central dipole 中心偶极子
central gradient array method 中间梯度法
central meridian 中央子午线
centrifugal drift 离心漂移
centrifugal force 离心力
chain reaction 链式反应
Chandler wobble 钱德勒摆动，钱德勒章动

channel 通道；航道
channel multiplier 通道倍增器
channel wave 通道波
Chapman layer 查普曼层
Chapman production function 查普曼生成函数
characteristic curve 特性曲线
characteristic wave 特征波
charge-exchange reaction 电荷交换反应
Charon 冥卫一
chart 海图
check station 检测台；监视台
checkup of seismic intensity 地震烈度复核
chemical abundance 化学丰度
chemical and dynamic lifetime 化学及动力学寿命
chemical cleaning 化学清洗
chemical equilibrium 化学平衡
chemical geothermometer 化学地球温度计
chemical remanent magnetization (CRM) 化学剩磁
chemiluminance 化学发光
chemopause 化学层顶
chemosphere 化学层
chirp 线性调频脉冲
chorisogram method 分区统计图表法
choroplethic map 等值区域图，分区量值地图
choroplethic method 分区统计图法
chromosphere 色球
chromospheric spicule 色球针状物
chronology 纪年法，年代学
Ci index Ci 指数
C index C 指数
circular curve location 圆曲线测设
circularly polarized 圆偏振的
Circum-Pacific Seismic Zone 环太平洋地震带
circumzenithal arc 环天顶弧

cirriform cloud 卷云状云
CIS (cartographic information system) 地图信息系统
cislunar space 地月空间
civil calendar 民用历
Clairaut theorem 克莱罗定理
classical auroral zone 古典极光带
clearance limit survey 净空区测量
cleft 极隙
climatological 气候(学)的
clinometer 倾斜仪
clipping 剪辑
clock correction 钟差
closed drift orbit 闭合漂移轨道
closed leveling line 闭合水准路线
closed loop field 闭合回线场
closed traverse 闭合导线
close-range photogrammetry 近景摄影测量
closing error 闭合差
closure 闭合(差)
cloud altimeter 云高计
CLVD (compensated linear vector dipole) 补偿线性向量偶极
CMB (core-mantle boundary) 核-幔边界
CN radical reaction CN基反应
coalescence of droplet 云滴并合
coal seismic prospecting 煤田地震勘探
Coarse/Acquisition Code (C/A Code) 粗码
coastal zone 海岸带
coasting flight 惯性飞行,滑翔飞行
coastline 海岸线
coastwise survey 沿岸测量,沿海测量
Cochiti event 科奇蒂事件
coda 尾波
codeclination 余赤纬,极距
co-geoid 共大地水准面
cognitive mapping 认知制图

coherence 相干
coherence emphasis 相干加强
coherence stack 相干叠加
coherent carrier 相干载波
coincidence counter 符合计数器
cold cathode 冷阴极
cold plasma 冷等离子体
cold plume 冷焰
cold spring 冷泉
collapse earthquake 陷落地震
collective effect 集体效应,集约效应
collimator 准直仪
collisional damping 碰撞阻尼
collision strength 碰撞强度
colour coding 彩色编码
colour coordinate system 彩色坐标系
colour enhancement 彩色增强
colour film 彩色片
colour infrared film 彩色红外片,假彩色片
colour management system 色彩管理系统
colour manuscript 彩色样图
colour photography 彩色摄影
colour proof 彩色校样
colour reproduction 彩色复制
colour sensitive material 彩色感光材料
colour separation 分色
colour space 颜色空间
colour transformation 彩色变换
colour wheel 色环
column density 柱密度
combined adjustment 联合平差
common-depth-point grid (CDP grid) 共深度点网格
common-depth-point stacking (CDP stacking) 共深度点叠加
common mid-point stacking 共中心点叠加
communication satellite 通信卫星
compact radio source 致密射电源

comparative cartography 比较地图学
comparison survey 联测比对
comparison with adjacent chart 邻图拼接比对
compass 罗盘仪
compass adjustment beacon 罗经标
compass survey 罗盘仪测量
compass theodolite 罗盘经纬仪
compensated linear vector dipole (CLVD) 补偿线性向量偶极
compensating error of compensators 补偿器补偿误差
compensation of undulation 波浪补偿
compensator 补偿器
compilation 编绘,编制
compiled original 编绘原图
complex correction 混合改正
complex ion 复离子
complex refractive index 复折射指数
complex resistivity method 复电阻率法
component 分量
component-based GIS 组件式 GIS
component of force 分力
composite fault-plane solution 综合断层面解
composite profiling method 联合剖面法
comprehensive atlas 综合地图集
comprehensive logging 综合录井
comprehensive map 综合地图
compressed magnetosphere 受压磁层
compressible fluid 可压缩流体
compressional wave 压缩波
compression of the earth 地球椭率
computer-aided cartography (CAC) 机助地图制图
computer-aided mapping 机助制图,机助测图

computer-assisted classification 机助分类
computer-assisted plotting 机助制图,机助测图
computer cartographic generalization 计算机制图综合
computer compatible tape (CCT) 计算机兼容磁带
computer vision 计算机视觉
conductive heat flow 传导热流
conductivity logging 电导率测井
conductivity tensor 电导率张量
confining pressure 围压
conformal projection 等角投影,正形投影
conglomerate test 砾石检验
conical wave 锥面波
conic projection 圆锥投影
coniferous forest 针叶林
conjugate photoelectron 共轭光电子
connate water 原生水
connection survey 联系测量
connection survey in mining panel 采区联系测量
connection triangle method 联系三角形法
Conrad discontinuity 康拉德界面
Conrad interface 康拉德界面
conservation law 守恒定律
conservation of mass 质量守恒
conserved quantity 守恒量
Consol chart 康索尔海图
constant altitude 恒定高度
constant phase relationship 常相位关系
construction land 建设用地
construction survey for shaft sinking 凿井施工测量
constructive interference 建设性干扰
contact induced polarization method 接触激发极化法
contact printing 接触印刷,接触晒印
continental drift 大陆漂移

continental fitting 大陆拼合
continental plate 大陆板块
continental reconstruction 大陆重建
continental splitting 大陆分裂
continental spreading 大陆扩张
continuation 延拓
continuation of potential field 位场延拓
continuous absorption 连续吸收
continuous attenuator 连续减光板
continuous emission 连续发射
continuous mode 连续方式
continuous spectrum 连续谱
continuous tone 连续调
contour 等高线
contourite mound seismic facies 等深流丘状地震相
contour line 等值线
contour map 等值线(图)
contour map migration 等值线图偏移
controlled source 可控震源
controlled source seismology 可控源地震学
control network for deformation observation 变形观测控制网
control point 控制点
control strip 测控条
control survey 控制测量
control survey of mining areas 矿区控制测量
convection 对流
convection cell 对流环
convection pattern 对流图型
convective heat flow 对流热流
convective region 对流区
convergence belt 汇聚带
convergence zone 汇聚带
convergent photography 交向摄影
convergent-type geothermal belt 汇聚型地热带
conversion 转换
conversion of waves 波的转换
conversion factor 转化因子

converted wave 转换波
cooling process 冷却过程
cool spring 凉泉
coordinate 坐标
coordinate momentum space 坐标动量空间
coordinate system 坐标系
coordinate transformation 坐标转换
coordinate universal time (UTC) 协调世界时
coplanar orbit 共面轨道
Cordelia 天卫六
core 岩心
core field 地核磁场
core-mantle boundary (CMB) 核-幔边界
core-mantle coupling 核-幔耦合
corner cube display 角视立体图
corner frequency 拐角频率
corona discharge 电晕放电
coronal disturbance 日冕扰动
coronal streamer 冕流
coronal wind 冕风
corotating magnetic field line 共旋磁力线
corotation field strength 共旋电场强度
corpuscular cosmic ray 微粒宇宙线
corpuscular eclipse 微粒食
corpuscular effect 微粒效应
corpuscular radiation 微粒辐射
corrected dipole coordinate 修正偶极坐标
corrected geomagnetic coordinate 修正地磁坐标
correction 校正,改正
correction for centring 归心改正
correction for deflection of the vertical 垂线偏差改正
correction for skew normals 标高差改正
correction from normal section to geodetic 截面差改正
correction of depth 测深改正

correction of transducer baseline 换能器基线改正
correction of transducer draft 换能器吃水改正
correction of zero drift 零漂改正
correction of zero line 零线改正
correlate 联系为
correlated fluctuation 相关涨落
correlator 相关器
co-seismic 同震的
cosine law 余弦定律
cosmic abundance 宇宙丰度
cosmical aerodynamics 宇宙空气动力学
cosmic background radiation 宇宙背景辐射
cosmic mapping 宇宙制图
cosmic noise 宇宙噪音
cosmic plasma 宇宙等离子体
cosmic radio noise 宇宙射电噪声
cosmic-ray abundance 宇宙线丰度
cosmic-ray equator 宇宙线赤道
cosmic-ray jet 宇宙线集流
cosmic-ray knee 宇宙线膝
cosmic-ray storm 宇宙线暴
cosmic X-radiation 宇宙X射线辐射
cosmochemistry 宇宙化学
cosmogony 天体演化学
Coulomb collision 库仑碰撞
Coulomb damping 库仑阻尼
Coulomb's law 库仑定律
coupled ion mass spectrometer 耦合离子质谱计
course 航向
covariance 协方差
covariance function 协方差函数
cover 覆被
Cowling conductivity 柯林电导率
creep 蠕变，蠕滑
creepmeter 蠕变仪
Cressida 天卫九
critical collision frequency 临界碰撞频率

critical frequency 临界频率
CRM (chemical remanent magnetization) 化学剩磁
crooked line seismic 弯线地震
cross correlation 互相关
cross-coupling effect 交叉耦合效应，C-C效应
cross-fault 跨断层
crosshole seismic 井间地震
crossline 联络测线
cross-modulation 交叉调制
cross-ruling 交叉网线
cross-section profile 横断面图
cross-section survey 横断面测量
crust 地壳
crustal 地壳的
crustal deformation 地壳形变
crustal earthquake 地壳地震
crustal transfer function 地壳传递函数
crystallization remanence 结晶剩磁
crystallization remanent magnetization 结晶剩磁
crystal spectrometer 晶体分光计
cultural map 文化地图
cumulative duration 累积持续时间
cumulus cloud line 积云线
Curie point 居里点
Curie temperature 居里温度
current electrode 供电电极
currently-used model atmosphere 现用标准大气
current surveying 测流
curtain shutter 焦面快门，帷幕式快门
curvature drift 曲率漂移
curvature vector 曲率矢量
cusp 尖点，极隙
cut-off rigidity 截止刚度
cyclic magnetization 循环磁化（强度）
cyclogenesis 气旋生成，气旋发生
cylindrical projection 圆柱投影
Czapski condition 交线条件，恰普斯基条件，向甫鲁条件

Dd

daily magnetic character figure 日磁情指数
dark flocculus 暗谱斑
dasymetric map 分区密度地图
data aquisition 数据采集
database design 数据库设计
database system 数据库系统
data capture 数据采集
data compression 数据压缩
data format 数据格式
data interpretation 数据解释
data processing 数据处理
data quality control 数据质量控制
data reduction 数据压缩,数据简化
data sample 数据样品
data standard 数据标准
data updating 数据更新
data visualization 数据可视化
data warehouse 数据仓库
datum 基准
datum static correction 基准面静校正
dawn and dusk meridian 晨昏子午线
dawn-dusk electric field 晨昏电场
dawn meridian 黎明子午线
day airglow 日气辉
daylight effect 昼光效应
DC cleaning (direct current cleaning) 直流清洗
DD model (dilatancy-diffusion model) 膨胀-扩散模式
deceleration radiation 减速辐射
declination 磁偏角

declination circle 赤纬圈
deconvolution 反褶积
decoupling 解耦
deep-focus earthquake 深(源地)震
deep seismic sounding 深地震测深
deep space 深空
deep space probe 深空探测器
deep water seismic 深海地震勘探
deflecting magnet 致偏磁体
deflection of the vertical 垂线偏差
deformation 形变
deformation monitoring 形变监测
degree of ionization 电离度
De Hoop transformation 德胡普变换
Deimos 火卫二
Delaunay triangulation Delaunay 三角网,德洛内三角网
delta 三角洲
DEM (digital elevation model) 数字高程模型
demultiplex 多路解编
density logging 密度测井
density of ionization 电离密度
density profile 密度剖面
depletion layer 耗尽层
depositional DRM 沉积碎屑剩磁
depositional remanence 沉积剩磁
depositional remanent magnetization (DRM) 沉积剩磁
depth 深度
depth migration 深度偏移
depth of compensation 补偿深度

depth perception 深度感
depth (record) section 深度剖面
Desdemona 天卫十
desert 荒漠,沙漠
desertification 荒漠化,沙漠化
design spectrum 设计谱
detectable solar wind 可测太阳风
detonating fireball 发声火流星
detrital magnetic particle 碎屑磁颗粒
detrital remanence 碎屑剩磁
detrital remanent magnetization (DRM) 碎屑剩磁
deuterium 氘,重氢
deuteron 氘核
deviational absorption 偏移吸收
dew point temperature 露点温度
diatomic ion 双原子离子
diatomic molecule 双原子分子
diazo copying 重氮复印
dielectric logging 介电测井
difference of latitude 纬度差
difference of longitude 经度差
difference threshold 差异阈
differential 微分
differential density spectrum 微分密度谱
differential energy spectrum 微分能谱
differential equation of geodesic 大地线微分方程
diffraction 绕射,衍射
diffuse solar radiation 太阳漫辐射
diffusion transfer 扩散转印
diffusive equilibrium 扩散平衡
digisonde 数字式测高仪
digital city 数字城市
digital elevation model (DEM) 数字高程模型
digital file 数字化文件
digital format 数据格式
digital graphic processing 数字图形处理
digital map 数字地图

Digital World Wide Standard Seismograph Network (DWWSSN) 数字化世界标准地震台网
digitization 数字化
dilatancy 膨胀
dilatancy-diffusion model (DD model) 膨胀-扩散模式
dilatancy hardening 膨胀硬化
dilatational wave 膨胀波
dilatometer 膨胀仪
dilute plasma 稀释等离子体
dim spot (地震勘探)暗点
Dione 土卫四
Dione B 土卫十二
dip 倾角
dip angle (磁)倾角
dip circle 磁倾圈
dip equator 倾角赤道
dipmeter survey 倾角测井
dip move-out (DMO) 倾斜时差校正
dipole 偶极(子)
dipole coordinate 偶极子坐标
dipole-dipole array 偶极排列
dipole-dipole array method 偶极排列法
dipole electrode array 偶极排列
dipole electrode sounding 偶极测深
dipole meridian 偶极子午线
dip orientation 倾向定向
dip pole 磁倾极
direct conductivity 直接电导率
direct current cleaning (DC cleaning) 直流清洗
direction cosine 方向余弦
direction of magnetization 磁化方向
directivity 方向性
directivity function 方向性函数
direct photoionization 直接光致电离
direct satellite probing 卫星直接探测
direct solar beam 直接太阳光束

direct solution of geodetic problem 大地主题正解
direct wave 直达波
disaster 灾害
disaster mitigation 减灾
disaster prediction 灾害预测
disaster prevention 防灾
discrete beam 不连续束
discrete polar cap aurora 离散极盖区极光
discrete spectrum 离散谱
discrete wavenumber-finite element method (DWFE method) 离散波数有限元法
discrete wavenumber method 离散波数法
dispersion relation 色散关系
dispersion wave 频散波
displacement response spectrum 位移反应谱
dissipation 耗散
dissociation 离解
dissociation potential 离解电势,离解电位
dissociative recombination 离解性复合
distance correction in Gauss projection 高斯投影距离改正
distant earthquake 远震
distorted wave 畸变波
distortion isogram 等变形线
distortion of projection 投影变形
distributed 分布式的
disturbance local-time inequality 扰动场地方时不均匀性
disturbance vector 扰动矢量
disturbed daily variation 扰日日变化
disturbing 扰动
disturbing mass 扰动质量
disturbing potential 扰动位
diurnal solar heating 周日太阳加热
divergence belt 发散带
divergence zone 发散带
diversity stack 花样叠加

diving wave 潜波
DMO (dip move-out) 倾斜时差校正
dominant wind 盛行风
Doppler frequency shift 多普勒频移
dot 网点
dot method 点值法
double nuclear resonance magnetometer 双重核共振磁力仪
double sunspot cycle 黑子双周
doughnut-shaped zone 轮胎形辐射带
down coming sky wave 下射天波
down sweep 降频扫描
draconic month 交点月
drag effect 曳引效应,曳力效应
D region D区
D region ion chemistry D区离子化学
D region negative ion D区负离子
drift 漂移
drift orbit 漂移轨道
drilling 钻进,钻探
drilling fluid 钻井液
drill seismic facies section 钻井-地震相剖面图
driving frequency 激励频率
DRM^1 (depositional remanent magnetization) 沉积剩磁
DRM^2 (detrital remanent magnetization) 碎屑剩磁
dropsonde 下投式探空仪
dry model 干模式
Dst index Dst指数
duct 导管
ducted propagation 导管传播
dune 沙丘
duration of shaking 震动持续时间
duration of totality 全食时间
DWFE method (discrete wavenumber-finite element method) 离散波数有限元法
DW method DW法

DWWSSN (Digital World Wide Standard Seismograph Network) 数字化世界标准地震台网
dye line proof 彩色线划校样
dynamical mechanical magnification 动态机械放大倍数
dynamic equalization 道内动平衡
dynamic geodesy 动力大地测量学
dynamic height 力高
dynamic landscape simulation 动态地景仿真
dynamic map 动态地图
dynamic range 动态范围
dynamic solar wind 动力太阳风
dynamic variable 动态变量
dynamo region 发电机区

earth 地球
earth crust structure 地壳构造
earth-flattening approximation 地球变平近似
earth-flattening transformation 地球变平换算
earth mantle 地幔
earth model 地球模型
earth point （陨石）触地点
earth pulsation 地脉动
earthquake 地震
earthquake catalogue 地震目录
earthquake damage 震害
earthquake depth 震源深度
earthquake disaster 地震灾害
earthquake dislocation 地震位错
earthquake engineering 地震工程（学）
earthquake force 地震力
earthquake forecasting 地震预报
earthquake frequency 地震频度
earthquake-generating stress 引震应力,发震应力
earthquake hazard 震灾
earthquake intensity 地震烈度
earthquake light 地震光
earthquake loading 地震载荷
earthquake location 地震定位
earthquake magnitude 震级
earthquake mechanism 地震机制
earthquake migration 地震迁移
earthquake period 地震周期
earthquake periodicity 地震周期性
earthquake prediction 地震预测
earthquake preparation 地震孕育,孕震

earthquake prevention 地震预防
earthquake-prone area 地震危险区
earthquake-proof 抗震的
earthquake province 地震区
earthquake recurrence rate 地震重复率
earthquake region 地震区
earthquake-resistant structure 抗震结构
earthquake risk 地震危险性
earthquake rupture mechanics 地震破裂力学
earthquake sequence 地震序列
earthquake series 地震系列
earthquake sound 地声
earthquake source mechanism 震源机制
earthquake statistics 地震统计（学）
earthquake swarm 震群
earthquake warning 地震警报
earthquake wave 地震波
earth resistivity 地电阻率
earth rotation 地球自转
earth spheroid 地扁球体
earth surface 地表
earth tide 固体潮,陆潮
earth tilt 地倾斜
eccentric dipole 偏心偶极子
eccentricity of ellipsoid 椭球偏心率
echo signal of sounder 测深仪回波信号
echo sounder 回声测深仪
echo sounding 回声测深
eclipse effect （日）食效应

eclipse of satellite 卫星食
eclipse of the moon 月食
eclipse of the sun 日食
ecliptic coordinate 黄道坐标
ecliptic plane 黄道面
economic map 经济地图
Eddington limit 爱丁顿极限
eddy energy 涡旋能量
edge detection 边缘检测
edge dislocation 刃型位错
edge enhancement 边缘增强
edge matching 图幅接边
edge of the format 图廓
effective atmospheric transmission 有效地球透射
effective atomic weight 有效原子量
effective peak acceleration (EPA) 有效峰值加速度
effective peak velocity (EPV) 有效峰值速度
effective radius of the earth 有效地球半径
effective stress 有效应力
effective terrestrial radiation 有效地球辐射
effective wave 有效波
effect of point 尖端效应
eigen value 本征值
Elara 木卫七
elastic collision 弹性碰撞
elastic rebound 弹性回跳
elastic wave 弹性波
electrical lateral curve logging 横向(电)测井
electrical logging 电测井
electrical prospecting 电法勘探
electrical sounding 电测深
electrical survey 电法调查
electric conductivity 电导率
electric polarization field 电极化场
electrode array 电极排列
electrode potential logging 电极电位测井
electro-dynamic effect 电动力效应
electrojet 电集流
electromagnetic 电磁的
electromagnetic field 电磁场
electromagnetic induction method 电磁感应法
electromagnetic method 电磁法
electromagnetic radiation 电磁辐射
electromagnetic seismograph 电磁式地震仪
electromagnetic sounding 电磁测深
electromagnetic vibration exciter 电磁脉冲震源
electromagnetic wave 电磁波
electron aurora 电子极光
electron chemistry 电子化学
electron density 电子密度
electron energy spectrum 电子能谱
electronic atlas 电子地图集
electronic map 电子地图
electronic publishing system 电子出版系统
electron zone 电子极光带
electrostatic equilibrium 静电平衡
element of absolute orientation 绝对定向元素
element of centring 归心元素
element of exterior orientation 像片外方位元素
element of interior orientation 像片内方位元素
element of rectification 纠正元素
element of relative orientation 相对定向元素
elevation 高程
elevation angle 高度角
ellipsoid 椭球
ellipsoidal geodesy 椭球面大地测量学
ellipsoidal height 大地高
ellipticity correction 椭率改正
emanation survey 射气测量

emersio 缓始
emission coefficient 发射系数
emission line 发射谱线
Enceladus 土卫二
end of totality 全食终
endogenous steam 内生蒸汽
end overlap 航向重叠
energy conversion efficiency 能量转化效率
energy spectrum 能谱
engineering seismology 工程地震（学）
engineering survey 工程测量
enhanced radiation 增强辐射
entropy wave 熵波
environmental biogeochemistry 环境生物地球化学
environmental chemistry 环境化学
environmental geochemistry 环境地球化学
environmental map 环境地图
environmental organic geochemistry 环境有机地球化学
environmental satellite 环境卫星
environmental survey satellite 环境探测卫星
Eotvos correction 厄特沃什改正，厄缶改正
EPA (effective peak acceleration) 有效峰值加速度
ephemeris 星历
epicentral distance 震中距
epicentre 震中
epicentre azimuth 震中方位角
epicentre distribution 震中分布
epicentre intensity 震中烈度
epicentre migration 震中迁移
epifocus 震中
epigenetic geochemical anomaly 后生地球化学异常
Epimetheus 土卫十一
epipolar correlation 核线相关
epipolar line 核线
epipolar plane 核面

epipolar ray 核线
epipole 核点
epoch 历元
EPV (effective peak velocity) 有效峰值速度
equal value gray scale 等值灰度尺
equation of radiative transfer 辐射转移方程
equator 赤道
equatorial 赤道的
equatorial absorption 赤道吸收
equatorial anomaly 赤道异常
equatorial electrojet 赤道电集流
equatorial ionosphere 赤道电离层
equatorial radius 赤道半径
equatorial satellite 赤道卫星
equidistant projection 等距投影
equilibrium energy spectrum 平衡能谱
equilibrium tide 平衡潮
equinoctial point 二分点
equipotential line 等位线
equipotential surface of gravity 重力等位面
equivalent air path 等效空气程
equivalent current system 等效电流系
equivalent projection 等积投影
equivoluminal wave 等体积波
E-region E区
erosion 侵蚀
error 误差
error source 误差来源
ERS (Europe Remote Sensing Satellite) 欧洲遥感卫星
eruptive arch 爆发拱
eruptive prominence 爆发日珥
escape flux 逃逸通量
ESP (extended seismic profiling) 扩展地震剖面法
Europa 木卫二
Europe Remote Sensing Satellite (ERS) 欧洲遥感卫星
evanescent wave 消散波

evaporation 蒸发量
evaporation process 蒸发过程
evapotranspiration 蒸散
evening sector 黄昏区段
evergreen broad-leaved forest 常绿阔叶林
excess ionization 过量电离
exchanging document of mining survey 矿山测量交换图
excitation-at-the-mass method 充电法
excited state 激发态
exobase 逸散层底
exobiology 外空生物学
exoenergic process 释能过程
exosphere 外逸层,逸散层
expansive phase 膨胀相
experimental geochemistry 实验地球化学
exploration 勘探,勘察,探测
exploration geochemistry 勘察地球化学
exploration geophysics 勘探地球物理(学)
exploration seismology 勘探地震学
explosion seismology 爆炸地震学
explosive cord 导炸索

explosive shower 爆发簇射
explosive source 爆炸震源
exposure 曝光
extended distance 延伸距离
extended seismic profiling (ESP) 扩展地震剖面法
extensional organization 整体结构
extensive coherent shower 广延相干簇射
extensometer 伸长仪
external electric field 外电场
external field 外源场
external magnetic field 外磁场
extinction coefficient 消光系数
extra contour 助曲线
extragalactic compact radio source 河外致密射电源
extragalactic cosmic ray proton 河外宇宙线质子
extraordinary wave 非寻常波
extrapolation 外推
extra-terrestrial origin 地外源
extra-terrestrial seismology 地外震学
extreme shallow water seismic 极浅海地震勘探
extreme ultraviolet line 极紫外线

F1 layer F1 层
F1 ledge F1 缘
F2 layer F2 层
fade 衰落
fade-out （短波通讯）中断
failure criterion 破坏准则
fair drawing 清绘
falling-sphere method 落球法
false colour composite 假彩色合成
false colour image 假彩色图像
false colour photography 假彩色摄影
false zodiacal light 假黄道光
family of comets 彗星族
Faraday effect 法拉第效应
Faraday rotation 法拉第旋转
far-field 远场
far-field body wave 远场体波
far-field surface wave 远场面波
far IR channel 远红外波道
farmland 农田
farmland soil 农田土壤
far ultraviolet spectrograph 远紫外光摄谱仪
fast particle impact 快粒子碰撞
fault 断层,断裂
fault basin 断陷盆地
fault block 断块
fault depression 断陷
fault earthquake 断层地震
faulting 断层(作用)
fault-plane solution 断层面解
fault zone 断裂带
feature code 特征码
feature code menu 特征码清单
felt earthquake 有感地震

Fermi acceleration 费米加速
Ferreros formula 菲列罗公式
fictitious graticule 经纬网格
fiducial mark 框标
fiducial point 基准点
field 野外
field-aligned current 场向电流
field-aligned irregularity 场向不规则结构
field deformation 场形变
field difference 场差
field distribution 场分布
field-free space 无磁场空间
field geological map 野外地质图
field mapping 野外填图
field of force 力场
field-reversal 场(致)反向
figure-ground discrimination 图形-背景辨别
figure of the earth 地球形状
filter 滤光片
filtering 滤波
filter photometer 单色光度计
final construction survey 竣工测量
final original 出版原图,制印原图
finite difference migration 有限差分偏移
finite moving source 有限移动源
finiteness correction 有限性校正
finiteness factor 有限性因子
finiteness transform 有限性变换
finite opening angle 有限张角
first cosmic velocity 第一宇宙速度
first motion 初动

first motion approximation 初动近似
first movement 初动
First Point of Aries 春分点
fishing chart 渔业用图
fishing haven 渔堰
fishing rock 渔礁
fishing stake 渔栅
fissionable material 裂变材料
fission product 裂变产物
fissure observation 裂缝观测
fitting 拟合
fixed satellite 对地静止卫星
fixed source field 定场场
fixed source method 定场法
flaps 分版原图
flare particle 耀斑粒子
flare surge 耀斑激浪
flash spectrum 闪光谱
flat-layer approximation 平层近似
flat spot 平点
flattening of ellipsoid 椭球扁率
flickering arc 闪变弧
flight characteristic 飞行特性
flight-path recovery 航迹恢复
flight plan of aerial photography 航摄计划
floating zenith telescope 浮动天顶仪
flood 洪水
flood discharge 泄洪;洪水流量
flood peak 洪峰
flood season 汛期
flora 植物区系
flow cross-section 过水断面,流截面
flow velocity 流速
flow velocity distribution 流速分布
fluctuating 脉动
fluid 流体
fluid mechanics 流体力学
fluid property 流体性质
flume 水槽
fluorescent map 荧光地图
flux-gate magnetometer 磁通门磁力仪
fluxmeter 磁通(量)计
focal depth 震源深度
focal dimension 震源尺度
focal force 震源力
focal length 焦距
focal mechanism 震源机制
focal mechanism solution 震源机制解
focal plane shutter 帘幕式快门
focal process 震源过程
focal sphere 震源球
focal volume 震源体积
focusing effect 会聚效应,聚焦效应
fold test 褶皱检验
following sunspot 后随黑子
footage measurement of workings 巷道验收测量
football mode 足球振型
forbidden transition 禁戒跃迁
forbidden zone boundary line 禁区界限
Forbush decrease 福布什下降
forced oscillation 受迫振荡
forcing function 强迫函数
fore-and-aft overlap 航向重叠
forensic seismology 法律地震学
foreshock 前震
forest 森林
forest basic map 林业基本图
forestry 林业
forest stand 林分
forest survey 林业测量
format 格式
formation 地层
formation mean direction 建造平均方向
formation mechanism 形成机制
formation pressure 地层压力
forward 正演
forward modeling 正演模拟
forward overlap 前向重叠,航向重叠
forward scatter 前向散射
fossil geothermal system 古地热系统

fossil magnetization 化石磁化（强度）
fossil water 古水
fountain effect 喷泉效应
four colour printing 四色印刷
Fourier analysis 傅里叶分析
fractal 分形
fracture 断裂,裂缝,裂隙
fracture criterion 破裂准则
fractured 裂缝性的
frame camera 框幅摄影机
frame of reference 参考系
free air anomaly 自由空气异常
free atmosphere 自由大气
free electron 自由电子
free-free absorption 自由-自由吸收
free molecular flow 自由分子流
free oscillation 自由振荡
free solar wind 自由太阳风
freezing nucleus 冻结核
F-region F区
frequency 频率
frequency assignment 频率分配
frequency band 频带,频段
frequency distribution 频率分布
frequency domain 频率域
frequency drift 频率漂移,频漂
frequency error 频率误差
frequency offset 频偏
frequency scale 频标
frequency sounding method 频率测深法
frequency spectrum 频谱
frequency-time pattern 频时图式
frequency-time record 频时图
frequency-wavenumber migration (F-W migration) 频率波数偏移
fringe region of atmosphere 大气边缘层
frontal activity 锋面活动
frontal position 锋面位置
full wave theory 全波理论
fully ionized plasma 完全电离等离子体
fumarole 喷气孔
fumarolic field 冒汽地面
fundamental astronomical point 基本天文点
fundamental frequency 基频
fuzzy classifier method 模糊分类法
fuzzy image 模糊影像
F-W migration (frequency-wavenumber migration) 频率波数偏移

galactic centre 银河中心
galactic coordinate 银河坐标
galactic cosmic ray 银河宇宙线
galactic plane 银道面
Galilean satellites 伽利略卫星
Galitzin seismograph 加利津地震仪
Ganymede 木卫三
gap 空区
garden gate suspension "花园门"悬挂法
gas exploder 气爆震源
gas ionization chamber 气体电离室
gas laser 气体激光器
gas multiplication factor 气体倍增因子
gas-phase titration 气相滴定
gate 闸门
Gauss epoch 高斯期
Gauss grid convergence 高斯平面子午线收敛角
Gaussian beam 高斯波束
Gaussian distribution 高斯分布
Gauss-Kruger projection 高斯-克吕格投影
Gauss mid-latitude formula 高斯中纬度公式
Gauss plane coordinate 高斯平面坐标
GDSN (Global Digital Seismograph Network) 全球数字地震台网
Gegenschein 对日照
general atlas 普通地图集
general chart 普通海图
general conjugacy 一般共轭性
generalized Ohm's law 广义欧姆定律
generalized ray 广义射线
generalized ray theory (GRT) 广义射线理论
general map 普通地图
general precession 总岁差
geocentric 地心的
geocentric coordinate 地心坐标
geochemical anomaly 地球化学异常
geochemical balance 地球化学平衡
geochemical barrier 地球化学障
geochemical behaviour 地球化学性状
geochemical character of elements 元素的地球化学性质
geochemical classification 地球化学分类
geochemical control 地球化学的控制
geochemical culmination 地球化学的积顶点
geochemical cycle of elements 地球化学的元素循环
geochemical detailed survey 地球化学详查
geochemical differentiation 地球化学分异酌
geochemical dispersion 地球化学分散
geochemical drainage reconnaissance 地球化学水系普查
geochemical drainage survey 水地球化学测量

geochemical ecology 地球化学生态学
geochemical endemic 地球化学地方病
geochemical environment 地球化学环境
geochemical exploration 地球化学勘探
geochemical facies 地球化学相
geochemical gas survey 气体地球化学测量
geochemical gradient 地球化学梯度
geochemical index 地球化学指数
geochemical indicator 地球化学指标
geochemical landscape 地球化学景观
geochemical leading element 地球化学标准元素
geochemical map 地球化学图
geochemical mapping 地球化学填图
geochemical migration of elements 地球化学的元素迁移
geochemical process 地球化学过程
geochemical profile 地球化学剖面
geochemical prospecting 地球化学探矿,地球化学勘探
geochemical province 地球化学者,地球化学区
geochemical recognition 地球化学普查
geochemical reconnaissance 地球化学踏勘
geochemical relief 地球化学地势
geochemical soil survey 土壤地球化学测量
geochemical surface 地球化学面
geochemical survey 地球化学勘探
geochemistry 地球化学
geochemistry of individual elements 个别元素地球化学
geochemistry of landscape 景观地球化学
geochemistry of lithosphere 岩石圈地球化学
geochemistry of mineral deposits 矿床地球化学
geochemistry of the atmosphere 大气地球化学
geochemistry of the heavy element 重元素地球化学
geochemistry of the hydrosphere 水圈地球化学
geochemistry of the soil 土壤地球化学
geochronology 地质年代学,地球纪年学
geocorona 地冕
geodesic 大地线
geodetic 大地测量的
geodetic astronomy 大地天文学
geodetic azimuth 大地方位角
geodetic coordinate 大地坐标
geodetic coordinate system 大地坐标系
geodetic datum 大地基准
geodetic height 大地高
geodetic latitude 大地纬度
geodetic longitude 大地经度
geodetic meridian 测地子午线
geodetic surveying 大地测量
geodynamics 地球动力学
geoelectric 地电的
geoelectric cross section 地电断面
geofluid 地热流体
geographical space 地理空间
geographic coordinate 地理坐标
geographic grid 地理格网
geographic information 地理信息
geographic information communication 地理信息传输
geography 地理学
geoheat 地热
geoid 大地水准面
geoisotherm 等地温面
geological 地质的

geological interpretation of photograph 像片地质判读,像片地质解译
geological map 地质(平面)图
geological photomap 影像地质图
geological radar 地质雷达
geomagnetic 地磁的
geomagnetic activity 地磁活动
geomagnetic axis 地磁轴
geomagnetic chronology 地磁年代学
geomagnetic coordinate 地磁坐标
geomagnetic coordinate system 地磁坐标系
geomagnetic cut-off momentum 地磁静止动量
geomagnetic disturbance activity 磁扰活动性
geomagnetic equator 地磁赤道
geomagnetic excursion 地磁漂移
geomagnetic field 地磁场
geomagnetic index 地磁指数
geomagnetic observation 地磁观测
geomagnetic physics 地磁物理学
geomagnetic polarity reversal 地磁极性反向
geomagnetic polarity reversal time scale 地磁极性转向年表
geomagnetic polarity time scale 地磁极性年表
geomagnetic pole 地磁极
geomagnetic response 地磁响应
geomagnetic station 地磁台
geomagnetic storm 磁暴
geomagnetic survey 地磁测量,磁测
geomagnetism 地磁(学)
geomatics 测绘学
geometric condition 几何条件
geometric correction 几何校正
geometric geodesy 几何大地测量学
geometric model 几何模型
geometric orientation 几何定向
geometric rectification 几何校正

geometric spreading 几何扩散
geometrisation of ore body 矿体几何制图
geomorphological map 地貌图
geomorphology 地貌学
geophone 检波器
geophone array 组合检波
geophysical 地球物理的
geophysical anomaly 地球物理异常
geophysical exploration 地球物理勘探,物探
geophysical field 地球物理场
geophysical method 地球物理方法
geophysical prospecting 地球物理勘探,物探
geophysical well-logging 地球物理测井
geophysics 地球物理学
geopotential 大地位
geopotential number 地球位数
geoscience 地学
geostationary satellite 同步卫星
geosynchronous altitude 同步高度
geotherm 地热
geothermal 地热的
geothermal activity 地热活动
geothermal anomaly 地热异常
geothermal energy 地热能
geothermal field 地热田
geothermal fluid 地热流体
geothermal gradient 地温梯度
geothermally-anomalous area 地热异常区
geothermal phenomenon 地热现象
geothermal prospecting 地热勘探
geothermal reservoir 地热水库
geothermal resources 地热资源
geothermal survey 地热调查
geothermal system 地热系统
geothermal water 地下热水
geothermics 地热学
geothermometer 地球温度计

geyser 间歇泉
geyserland 间歇泉区
ghost reflection 虚反射
Gilbert epoch 吉尔伯特期
Gilbert reversed polarity epoch 吉尔伯特反极性期
Gilsa event 吉尔绍事件
GIS technology 地理信息系统技术
glacial epoch 冰川时代
glacier 冰川
glaciology 冰川学
global atmospheric circulation 全球大气环流
global circulation 全球环流
Global Digital Seismograph Network (GDSN) 全球数字地震台网
global heat balance 全球热平衡
global heat budget 全球热量收支
global meteorology 全球气象学
Global Navigation Satellite System (GLONASS) 全球导航卫星系统
global picture 全球图
Global Position System (GPS) 全球定位系统
global wind system 全球风系
globe 地球仪
GLONASS (Global Navigation Satellite System) 全球导航卫星系统
glow cloud experiment 发光云实验
gnomonic projection 球心投影,日晷投影,大环投影
GPS (Global Position System) 全球定位系统
GPS-based 基于GPS的
GPS control network GPS控制网
GPS network GPS网
GPS observation GPS观测
GPS positioning GPS定位
GPS receiver GPS接收机
GPS satellite GPS卫星
GPS signal GPS信号
GPS survey GPS测量
GPS technology GPS技术
grade location 坡度测设
gradient drift current 梯度漂移电流
gradient wind 梯度风
gradual commencement (magnetic) storm 缓始磁暴
graduation of tints 分层设色表
Graefenberg array 格拉芬堡台阵
Graham magnetic interval 格拉姆磁间段
graphic element 图形元素
graphic sign 图形记号
graphic symbol 图形符号
grassland 草地
grating spectrograph 光栅摄谱仪
grating spectrum 光栅光谱
gravimeter 重力仪,重力计
gravimeter drift correction 重力仪零漂改正
gravimetry 重力测量学
gravitation 引力
gravitational interaction 引力相互作用
gravitational potential 引力位
gravitational radiation 引力辐射
gravitational red shift 引力红移
gravitational tide 引力潮
gravity 重力
gravity acceleration 重力加速度
gravity anomaly 重力异常
gravity anomaly due to magnetic body 磁源重力异常
gravity field 重力场
gravity gradient survey 重力梯度测量
gravity gradient zone 重力梯度带
gravity gradiometer 重力梯度仪
gravity high 重力高
gravity low 重力低
gravity maximum 重力高
gravity measurement 重力测量
gravity measurement at sea 海洋重力测量

gravity minimum 重力低
gravity potential 重力位
gravity prospecting 重力勘探
gravity survey 重力调查
gray atmosphere 灰体大气
grazing angle 掠射角
grazing incidence 掠入角
Great Red Spot 大红斑
greenhouse effect 温室效应
grey wedge 灰楔
grid 网格
grid bearing 坐标方位角
grid bin 网格单元
grid cell 网格单元
gridded chamber 加栅电离室
grid map 网格地图
grid method 网格法
grid of neighbouring zone 邻带方里网
grid structure 网格结构
grid variation 磁偏角
gross error 粗差
gross error detection 粗差检测
ground-backscatter 地面反向散射

ground-based measurement 地面测量
ground follow-up 地面查证
ground motion 地面运动
ground penetrating radar 探地雷达, 地质雷达
ground roll 地滚
ground state 基态
ground subsidence 地面沉降
ground tilt 地倾斜
groundwater 地下水
groundwater level 地下水位
ground wave 地表波
group of comets 彗星群
group velocity 群速度
GRT (generalized ray theory) 广义射线理论
Gruber point 标准配置点
guiding centre plasma 引导中心等离子体
gully 沟壑
gyrate clockwise 顺时针回旋
gyro horizon 陀螺地平仪
gyromagnetic 陀螺磁罗经的
gyroresonance 回旋共振

hachuring 晕瀚法
Hadamard transformation 阿达马变换
hair-pin curve location 回头曲线测设
half-interval contour 间曲线,半距等高线
half-power point 半功率点
half-shadow 半影
halftone 半色调
Hall conductivity 霍尔电导率
hammada 石漠,岩漠
hanging theodolite 悬式经纬仪
Harang discontinuity 哈朗间断
harder spectrum 较硬能谱
hardness of the spectrum 能谱硬性
hard radiation 硬辐射
harmonic analysis 谐波分析,调和分析
Hayford ellipsoid 海福德椭球
HCI (hydro carbon indicator) 烃类检测
head wave 首波
healing front 愈合前沿
heat budget 热量收支
heat-conduction equation 热传导方程
heat death 热寂
heat flow 热流
heat flow province 热流区
heat flow subprovince 热流亚区
heat flow unit (HFU) 热流单位
heat generation unit 生热率单位
heating of the plasma 等离子体加热
heating source experiment 热源实验
heat sink 热壑
heat source 热源
heat-source characteristics 热源特性
heave compensation 波浪补偿
heave compensator 波浪补偿器,涌浪滤波器
heavy charged particle 重带电粒子
heavy ion 重离子
heavy water 重水
height 高程
height interval 高度间隔
height measurement 高程测量
height of aurora 极光高度
heliocentric coordinate 日心坐标
heliocentric distance 日心距离
heliographic chart 日面图
heliographic coordinate 日面坐标
helioscope 太阳望远镜
heliosphere 日球层
heliospheric current sheet 日球层电流片
helium ion 氦离子
heterogeneity 非均质性,异质性
heterosphere 非均匀层
HFU (heat flow unit) 热流单位
hierarchical organization 等级结构
high-altitude nuclear explosion 高空核爆炸
high atmosphere wind 高空风
high drag region 高曳力区
high energy particle 高能粒子
higher mode 高阶振型
higher-order approximation 高阶近似

highest normal high water 理论最高高潮面
high frequency 高频
high ionospheric wind 高电离层风
high-latitude precipitation region 高纬沉降区
high-latitude spot 高纬黑子
highly elongated orbit 大扁度轨道
high precision 高精度
high pressure system 高气压系统
high resolution 高分辨率
high-speed plasma 高速等离子体
hill 丘陵
hill shading 晕渲法
Himalia 木卫六
historical baseline 历史基线
historical earthquake 历史地震
historical geochemistry 历史地球化学
historical seismology 历史地震学
historic map 历史地图
HLEM (horizontal loop method) 水平回线法
holocene 全新世
hologram photography 全息摄影
homeotheric map 组合地图
homing beacon 归航信标
homogeneous plasma 均匀等离子体
homomorphic deconvolution 同态反褶积
homopause 均匀层顶
homosphere 均匀层
horizon 层位
horizon flattening 层位拉平
horizontal component 水平分量
horizontal control network 平面控制网,水平控制网
horizontal control point 平面控制点
horizontal coordinate 平面坐标
horizontal diffusion 水平扩散
horizontal intensity 水平强度

horizontal loop method (HLEM) 水平回线法
horizontal refraction error 水平折光差,旁折光差
horizontal system of coordinate 地平坐标系
horizontal well 水平井
horizon trace 像地平线
hot dry rock 干热岩体
hot electron gas 热电子气体
hot magnetospheric plasma 磁层热等离子体
hot plasma ion 热等离子体离子
hot plume 热焰
hot spot 热点
Hough function 霍夫函数
hue 色相
human map 人文地图
humus 腐殖质
hurricane cloud 飓风云
hybrid resonance 混合共振
hydrate 水合物
hydraulic 水力的
hydraulics 水力学
hydrocarbon 烃,碳氢化合物
hydrocarbon indicator (HCI) 烃类检测
hydrodynamic 水动力的
hydrofracturing 水压致裂
hydrogen flocculus 氢谱斑
hydrogenous material 含氢物质
hydrogen peroxide 过氧化氢
hydrological 水文的
hydrological forecasting 水文预报
hydrological model 水文模型
hydrological process 水文过程
hydrologic station 水文站
hydrology 水文学
hydromagnetics 磁流体力学
hydroperoxyl radical reaction 氢过氧自由基反应
hydropower station 水电站
hydrostatic equilibrium 流体静力学平衡
hydrothermal activity 水热活动

hydrothermal alteration 水热蚀变
hydrothermal area 水热区
hydrothermal circulation 水热循环
hydrothermal convection system 水热对流系统
hydrothermal eruption 水热喷发
hydrothermal explosion 水热爆炸
hydrothermal field 水热田
hydrothermal mineralization 水热矿化
hydrothermal resource 水热资源
hydrothermal system 水热系统

hydroxyl 羟基
hyperbolic comet 双曲线轨道彗星
hyperfocal distance 超焦点距离
Hyperion 土卫七
hypocentral distance 震源距
hypocentral location 震源定位
hypocentre 震源
hypocentre parameter 震源参数
hypothesis of geochemical accumulators 地球化学蓄电池假说
hypsometric layer 分层设色法
hypsometric map 地势图
hysteresis effect 滞后效应

Iapetus 土卫八
IAT (International Atomic Time) 国际原子时
IAU (International Astronomical Union) 国际天文联合会
ice age 冰期
IDA Network (International Deployment of Accelerometers Network) 埃达台网,国际加速度计部署台网
identification code 识别码
identification of seismic events 地震波识别
igneous 火成的
igneous rock 火成岩
ignorosphere 未知层
Illawarra reversal 伊勒瓦拉反向
image 影像,图像
image correlation 影像相关
image data 影像数据,图像数据
image database 影像数据库,图像数据库
image dipole 像偶极子
image fusion 影像融合
image horizon 像地平线
image matching 影像匹配,图像匹配
image mosaic 影像镶嵌,图像镶嵌
image motion compensation (IMC) 像移补偿
image pyramid 影像金字塔
image quality 影像质量,图像质量
image resolution 影像分辨率,图像分辨率
image restoration 影像复原,图像复原
image setter 激光照排机
image space coordinate system 像空间坐标系
imaginary-real component method 虚实分量法
imaging 成像
imaging radar 成像雷达
imaging spectrometer 成像光谱仪
IMC (image motion compensation) 像移补偿
IMF (interplanetary magnetic field) 行星际磁场
impact acceleration 冲击加速度
impact zone 冲击带
impedance 阻抗
impedance interface 阻抗界面
impedance probe 阻抗探针
impending earthquake 临震
impulse heating 脉冲加热
IMS Bulletin 国际磁层研究公报
IMS Satellite Situation Centre 国际磁层研究卫星形势中心
IMS Steering Committee 国际磁层研究干事会
incidental prominence 偶现日珥
incident stream 入射流
inclination (磁)倾角
inclinometer 磁倾计
incoherent scattering 非相干散射
incoherent scattering radar 非相干散射雷达
increased dawn-dusk electric field 增强的晨昏电场
index contour 计曲线
index diagram 图幅接合表

index for selection 选取指标
index mosaic 镶嵌索引图
index of refraction 折射率
index of solar ultraviolet activity 太阳活动紫外指数
Indian spring low water 印度大潮低潮面
indicatrix ellipse 变形椭圆
indirect aerological analysis 间接高空分析
indirect scheme of digital rectification 间接法纠正
induced earthquake 诱发地震
induced polarization method (IP method) 激发极化法
induced pulse transient method (INPUT method) 感应脉冲瞬变法,因普特法
induced seismicity 诱发地震活动性
induction drag 感生阻力
induction height survey through shaft 立井导入高程测量
induction logging 感应测井
inert gas 惰性气体
inertia drift 惯性漂移
inertial axis 惯性轴
inertial coordinate system 惯性坐标系
inertial navigation 惯性导航
inferior planet 内行星
infiltration 入渗
information attribute 信息属性
information database 信息数据库
information extraction 信息提取
infrared astronomy 红外天文学
infrared atmospheric band system 红外大气光谱带系
infrared brightness temperature 红外亮温
infrared EDM instrument 红外测距仪
infrared film 红外片
infrared flux 红外辐射通量
infrared imagery 红外图像

infrared photography 红外摄影
infrared radiometer 红外辐射计
infrared remote sensing 红外遥感
infrared scanner 红外扫描仪
inhomogeneous 不均匀的
inhomogeneous medium 不均匀介质
initial disturbance 初始扰动
initial phase 初相
initial stress 初始应力
injection boundary 注入边界
inline 主测线
inner auroral zone 内极光带
inner-core 内核
inner planet 内行星
inner radiation belt 内辐射带
inorganic 无机的
inorganic geochemistry 无机地球化学
inorganic scintillator 无机闪烁体;无机闪烁器
in-plane shear crack 平面剪切裂纹
INPUT method (induced pulse transient method) 感应脉冲瞬变法,因普特法
inseam seismic method 槽波地震法
in situ measurement 原地测量
in situ probe 原地探针
in situ stress 原地应力
instantaneous latitude 瞬时纬度
instantaneous longitude 瞬时经度
instantaneous magnetic zenith 瞬时磁天顶
instantaneous pole 瞬时极
instrument of surveying and mapping 测绘仪器
integer ambiguity 整周模糊度
integral density spectrum 积分密度谱
integral effective spectrum 有效积分谱

integral electron concentration 积分电子浓度
integral flux 积分通量
integrated blackbody photon flux 黑体光子积分通量
integrated geodesy 整体大地测量
integrated geophysical system 综合物探系统
integrated navigation 组合导航
integrated radiation 累积辐射
integrated spectrum 积分谱
intense burst 强爆发
intensity 烈度
intensity scale 烈度表
interacting prominence 互扰日珥
interactive processing 人机交互处理
interchange instability 交换不稳定性
intercloud medium 云际介质
interference filter 干涉滤光片，干涉滤波器
interferometric seabed inspection sonar 相干声呐测深系统
intergalactic gas 星系际气体
interior orientation 内部定向
interlocking point 互换点
intermediate contour 首曲线
intermediate-energy particle 中能粒子
intermediate polarity 中间极性
intermittent spring 间歇泉
internal conversion 内转换
internal field 内源场
International Astronomical Union (IAU) 国际天文联合会
International Atomic Time (IAT) 国际原子时
International Aurora Atlas 国际极光图
International Deployment of Accelerometers Network (IDA Network) 埃达台网，国际加速度计部署台网
international ellipsoid 国际椭球体
international geomagnetic reference field 国际参考地磁场
International Geophysical Year 国际地球物理年
international magnetic character figure 国际磁情记数
international most disturbed days 国际最扰日
International Polar Year 国际极年
International Quiet Sun Year 国际宁静太阳年
international reference atmosphere 国际参考大气
international reference ellipsoid 国际参考椭球
international reference ionosphere 国际参考电离层
international reference model 国际参考模式
International Seismological Center (ISC) 国际地震中心
International Seismological Summary (ISS) 国际地震汇编
International System of Units 国际单位制
International Union of Surveying and Mapping 国际测绘联合会
interoperability 互操作
interplanetary 行星际的
interplanetary background 行星际背景
interplanetary current 行星际电流
interplanetary discontinuity 行星际间断
interplanetary disturbance 行星际扰动
interplanetary dust 行星际尘埃
interplanetary event 行星际事件
interplanetary magnetic field (IMF) 行星际磁场
interplanetary magnetic field line 行星际磁力线
interplanetary matter 行星际物质

interplanetary plasma 行星际等离子体
interplanetary scintillation 行星际闪烁
interplanetary shock 行星际激波
interplanetary solar plasma 星际太阳等离子体
interplanetary space 行星际空间
interplate earthquake 板间地震
interplate geothermal belt 板间地热带
interpolation 插值
interpretation 判读
interpretoscope 判读仪
interrupted projection 分瓣投影
intersection 交会
interstellar line 星际谱线
interstellar medium 星际介质
interstellar space （恒）星际空间
interstellar wind 星际风
interval scaling 等距量表
interval velocity 间隔速度，层速度
intervening layer 层间层
intraplate earthquake 板内地震
intraplate geothermal system 板内地热系统
intraplate volcano 板内火山
intrinsic brightness 本身亮度
intrinsic magnetic element 内禀磁要素
invar baseline wire 因瓦基线尺
invariant geomagnetic coordinate 不变地磁坐标
invariant latitude 不变纬度
invasion 侵入
inverse correlation 负相关
inverse dispersion 反频散
inverse solution of geodetic problem 大地主题反解
inversion 回返，反演
inversion layer 逆温层
inverted-V event 倒 V 事件
Io 木卫一
ion-acoustic instability 离子声不稳定性
ion-atom rearrangement 离子-原子重新排列
ion chemistry 离子化学
ion cluster 离子团
ion concentration 离子浓度
ion density 离子密度
ion-ion neutralization 离子-离子中和
ionization by collision 碰撞电离
ionization energy 电离能
ionization rate 电离率
ionized atmosphere 电离大气
ionized region 电离区
ionizing radiation 电离辐射
ion mass spectrometer 离子质谱计
ion mobility 离子迁移率
ion-molecule kinetics 离子-分子动力学
ionogram 电离图
ionopause 电离层顶
ionosonde 电离层测高仪
ionosphere 电离层
ionospheric 电离层的
ionospheric absorption 电离层吸收
ionospheric disturbance 电离层扰动
ionospheric D-region 电离层 D 区
ionospheric eclipse 电离层食
ionospheric modification 电离层改造
ionospheric plasma 电离层等离子体
ionospheric profile 电离层剖面
ionospherics 电文学
ionospheric storm 电离层暴
ion pair 离子对
ion wave 离子波
IP method (induced polarization method) 激发极化法
IRM (isothermal remanent magnetization) 等温剩磁
irregular auroral patch 不规则极光斑

irregular geomagnetic-field 不规则地磁场
irrigation 灌溉
irrotational wave 无旋波
ISC (International Seismological Center) 国际地震中心
island arc geothermal zone 岛弧地热带
isoabsorption line 等吸收线
isocentre of photograph 像等角点
isogeotherm 等地温面,等地温线
isoline map 等值线地图
isoline method 等值线法
isomagnetic chart 等磁图
isomagnetic line 等磁强线
isometric latitude 等量纬度
isopiestics 等压线
isopore 等年变线
isoporic line 等年变线

isoseismal curve 等震线
isoseismal line 等震线
isostasy 地壳均衡(说)
isostatic anomaly 均衡异常
isothermal layer 等温层
isothermal remanent magnetization (IRM) 等温剩磁
isotope 同位素
isotope geochemistry 同位素地球化学
isotope tracer 同位素示踪物
isotopic geothermometer 同位素地球温度计
isotropic flux 各向同性通量
isotropic particle 各向同性粒子
ISS (International Seismological Summary) 国际地震汇编
iteration method with variable weights 选权迭代法

Jalamillo event 哈拉米略事件
Janus 土卫十
Japan Meteorological Agency (intensity) scale 日本气象厅(烈度)表,JMA 表
Jeans instability criterion 琼斯不稳定判据
Jeans length 琼斯长度
Jeffreys-Bullen travel time table 杰弗里斯-布伦走时表
jet stream intensity 急流强度
jet stream zone 急流带
JMA (intensity) scale 日本气象厅(烈度)表,JMA 表
JND (just noticeable difference) 恰可察觉差

Johnson noise 约翰逊噪音
joint hypocentral determination 联合震源定位
Joule dissipation 焦耳耗散
Joule heating 焦耳加热
Jovian planet 类木行星
Juliet 天卫十一
Jupiter 木星
Jupiter cloud belt 木星云带
Jupiter heat radiation 木星热辐射
Jupiter red spot 木星红斑
just noticeable difference (JND) 恰可察觉差
juvenile gas 原生气体
juvenile water 原生水

Kaena event 卡埃纳事件
Kalman filter 卡尔曼滤波器
karst 喀斯特
kataseism 向源震
kataseismic onset 向源初动
key bed 标准层，标志层
Kiaman interval 基亚曼间段
kilometer grid 方里网
kilometer scale 千米尺
K index K 指数
kinetic energy density 动能密度

kinetic energy transformation 动能转换
kink instability 扭曲不稳定性
Kirchhoff integration migration 基尔霍夫积分偏移
K parameter K 参数
K precision parameter K 精度参数
Kp index Kp 指数
Krasovsky ellipsoid 克拉索夫斯基椭球

laboratory plasma 实验室等离子体
LaCoste suspension 拉科斯特悬挂法
lake survey 湖泊测量
Lambert projection 兰勃特投影,兰勃特等面积方位投影
lamellar domain 层状畴
land 国土;陆地
Landau damping 朗道阻尼
land cover 土地覆被
land desertification 土地沙漠化
landform 地貌
land price 地价
land resource 土地资源
Landsat 陆地卫星
landscape 景观
landscape ecology 景观生态学
landscape geochemical prospecting 景观地球化学勘探
landscape map 景观地图
landscape pattern 景观格局
landscape space 景观空间
landscape type 景观类型
landslide 滑坡
land use 土地利用
lane 相位周,巷
lane width 相位周值,巷宽
Laplace azimuth 拉普拉斯方位角
Laplace point 拉普拉斯点
large amplitude theory 大振幅理论
large-aperture seismic array (LASA) 大孔径地震台阵
Large Earthquake Prompt Report Network 大震速报台网
large-scale convergence 大尺度辐合
large-scale planetary wave 大尺度行星波
large-scale solar magnetic field 大尺度太阳磁场
Larmor frequency 拉莫尔频率
Larmor motion 拉莫尔运动
LASA (large-aperture seismic array) 大孔径地震台阵
Laschamp event 拉尚事件
Laschamp excursion 拉尚漂移
laser aligner 激光准直仪
laser altimeter 激光测高仪
laser diode (LD) 激光二极管
laser eyepiece 激光目镜
laser guide of vertical shaft 立井激光指向(法)
laser level 激光水准仪
laser plotter 激光绘图机
laser plumbing 激光投点
laser ranger 激光测距仪
laser sounder 激光测深仪
laser swinger 激光扫平仪
laser theodolite 激光经纬仪
laser topographic position finder 激光地形仪
late glacial time 晚冰期
lateral 横向的,侧面的
lateral overlap 旁向重叠
lateral tilt 旁向倾角
lateral wave 侧面波
laterolog 侧向测井
latitude 纬度
latitude correction 纬度校正
latitude effect 纬度效应
latitude-local time distribution 纬度-地方时分布

latitude of pedal 底点纬度
latitude of reference 基准纬度
launch azimuth 发射方位角
law of areas 面积定律
law of planetary distance 行星距离定律
laws of photochemistry 光化学定律
layer 分层
layered 层状的
LD (laser diode) 激光二极管
leading sunspot 前导黑子
leaking mode 泄漏振型
left-handed rotation 左旋
legend 图例
lens seismic facies 透镜状地震相
leveling 水准
leveling line 水准路线
leveling measurement 水准测量
leveling of model 模型置平
level of escape 逃逸高度
level surface 水准面
lifeline 生命线
lifetime of a satellite 卫星寿命
light element 轻元素
light ions 轻离子
lightness 亮度
lightning activity 闪电活动
light pressure 光压
light year 光年
limb darkening 临边昏暗
limit error 极限误差
linear-angular intersection 边角交会法
linear array sensor 线阵遥感器
linear intersection 边交会法
linear polarization 线偏振
linear triangulation network 线形网
line shift 谱线位移
line smoothing 曲线光滑
line splitting 谱线分裂
line symbol 线状符号
liquefaction 液化
liquid core 液核
lithium reaction 锂反应

lithology 岩性,岩石学
lithosphere 岩石层,岩石圈
litter 凋落物,枯落物
LLR (lunar laser ranging) 激光测月
load Love's number 载荷勒夫数
load tide 载荷潮
local anomaly 局部异常
local apparent noon 地方视正午
local apparent time 地方视时
local atmosphere 局部大气
local earthquake 地方震
localized heating 局部增温
local K index 区域K指数
local magnitude 地方震级
local meridian 当地子午线
local shock 地方震
local sidereal time 地方恒星时
local time distribution 地方时分布
location of pier 桥墩定位
locked fault 闭锁断层
loess 黄土
loess hill 黄土丘陵
loess hilly region 黄土丘陵区
loess plateau 黄土高原
log 测井图
logging 测井
logging curve 测井曲线
logging data 测井数据
logging technology 测井技术
logging tool 测井仪
logical consistency 逻辑兼容,逻辑一致性
longitude effect 经度效应
longitudinal conductance 纵向电导
longitudinal energy 纵向能量
longitudinal extent 经向范围
longitudinal overlap 航向重叠
longitudinal tilt 航向倾角
longitudinal wave 纵波
long-period circulation 长周期循环
long-range navigation 远程导航
long-range positioning system 远程定位系统

long-term average 长期均值
long-term variation 长期变化
long-wave radiation 长波辐射
loop prominence 圈状日珥
loop-shaped sounding 环形测深
Loran chart 罗兰海图
Loran-C positioning system 罗兰-C定位系统
loss cone 损失锥,逃逸锥
Love's number 勒夫数
Love wave 勒夫波
low-altitude rocket vehicle 低空火箭
low-energy electron 低能电子
low-energy particle 低能粒子
lower atmosphere 低层大气
lower crust 下地壳
lower ionosphere 低电离层
lower low water 略最低低潮面
lower mantle 下地幔
lower thermosphere 低热层
lowest normal low water 理论最低低潮面
low-latitude aurora 低纬极光
low-latitude storm 低纬磁暴
low-orbiting satellite 低轨卫星

low oxygen abundance 低氧丰度
low velocity layer (LVL) 低速层
low velocity zone (LVZ) 低速区
low water 枯水,低水位
loxygen 液氧
luminescence spectrum 发光谱
luminosity curve 光度曲线
luminous efficiency 发光效率
luminous flux density 光通量密度
luminous region 发光区
lunar crater 月面环形山
lunar eclipse 月食
lunar geodesy 月面测量学,月球测绘
lunar laser ranging (LLR) 激光测月
lunar orbiter 月球轨道飞行器
lunar oscillation 月球振荡
lunar seismogram 月震图
lunar seismology 月震学
lunar tide 太阴潮
lunisolar precession 日月岁差
LVL (low velocity layer) 低速层
LVZ (low velocity zone) 低速区
Lyman line 莱曼线

Mach number 马赫数
macrophotogrammetry 超近摄影测量
macroseismic data 宏观地震资料
magmatic chamber 岩浆房
magmatic circulation 岩浆环流
magmatic pocket 岩浆房
magmatic water 岩浆水
magnetically disturbed day 磁扰日
magnetically quiet day 磁静日
magnetic anomaly area 磁力异常区
magnetic azimuth 磁方位角
magnetic bay 磁湾扰
magnetic bearing 磁象限角
magnetic character figure 磁情记数
magnetic charging method 磁充电法
magnetic chart 磁图
magnetic cleaning 磁清洗
magnetic colatitude 磁余纬
magnetic configuration 磁位形
magnetic conjugate point 磁共轭点
magnetic coordinate 磁坐标
magnetic coupling 磁耦合
magnetic crochet 磁钩扰
magnetic declination 磁偏角
magnetic dip 磁倾角
magnetic dip angle 磁倾角
magnetic dipole 磁偶极
magnetic dipole time 磁偶极时
magnetic disturbance 磁扰
magnetic effect 磁效应
magnetic element 地磁要素,磁元

magnetic fabric 磁组构
magnetic field 磁场
magnetic field change 磁场变化
magnetic field polarity 磁场极性
magnetic figure 磁力线图
magnetic flux density 磁通量密度
magnetic gradient drift 磁场梯度漂移
magnetic gradiometer 磁力梯度仪
magnetic induced polarization method (MIP method) 磁激发极化法
magnetic isoanomalous line 等磁异常线
magnetic isoclinic line 等磁倾线
magnetic local time 磁地方时
magnetic meridian 磁子午线
magnetic micropulsation 地磁微脉动
magnetic moment 磁矩
magnetic overprinting 磁叠印
magnetic pressure 磁压
magnetic prospecting 磁法勘探
magnetic pumping process 磁泵过程
magnetic quiet zone 磁静带
magnetic rigidity 磁刚度
magnetic sector structure 磁扇形结构
magnetic shell 磁壳
magnetic signature 磁异常特征
magnetic sounder 磁测深线;磁测深仪
magnetic sounding 磁测深
magnetic spectrograph 磁谱仪

magnetic storm 磁暴
magnetic stratigraphy 磁性地层学
magnetic substorm 磁亚暴
magnetic survey 磁法调查
magnetic susceptibility logging 磁化率测井
magnetic susceptibility meter 磁化率计
magnetic sweeping 磁力扫海测量
magnetic variation 磁偏角
magnetic washing 磁清洗
magnetocrystalline anisotropy 磁晶各向异性
magnetogram 磁照图
magneto-ionic theory 磁离子理论
magnetometer 磁力仪
magnetopause 磁层顶
magnetosheath 磁鞘
magnetosheath region 磁鞘区
magnetosphere 磁层
magnetospheric convection 磁层对流
magnetospheric dumping 磁层倾泻
magnetospheric plasma 磁层等离子体
magnetospheric response 磁层响应
magnetospheric storm 磁层暴
magnetostratigraphy 磁性地层学
magnetotail 磁尾
magnetotelluric 大地电磁的
magnetotelluric method 磁大地电流法
magnetotelluric sounding 大地电磁测深
magnitude 地震震级
magnitude-frequency relation 震级-频度关系
main field 主磁场
main phase 主相
main shock 主震
main spot 主黑子
main zone 主带

major disturbance 大扰动
major earthquake 大震
major planet 大行星
major radius of ellipsoid 椭球长半径
Mammoth event 马默思事件
man-made dust 人工尘埃
mantle 地幔
mantle convection 地幔对流
mantle convection cell 地幔对流环
mantle heat flow 地幔热流
mantle plume 地幔焰,地幔柱
map border 图廓
map clarity 地图清晰性
map colour atlas 地图色谱
map colour standard 地图色标
map compilation 地图编制
map complexity 地图复杂性
map data structure 地图数据结构
map decoration 地图整饰
map design 地图设计
map digitizing 地图数字化
map display 地图显示
map editing 地图编辑
map editorial policy 地图编辑大纲
map interpretation 地图判读
map layout 图面配置
map legibility 地图易读性
map load 地图负载量
map making 地图制图
map of mineral deposits 矿产图
map of mining subsidence 开采沉陷图
map overlay analysis 地图叠置分析
map perception 地图感受
mapping 测图,制图
mapping method with transit 经纬仪测绘法
mapping recorded file 图历簿
mapping satellite 测图卫星
map printing 地图印刷
map projection 地图投影

map reproduction 地图复制
map revision 地图更新
mapsheet 图幅
map symbols bank 地图符号库
map use 地图利用
marine geochemical prospecting 海洋地球化学探矿
marine geodesy 海洋大地测量学
marine reflection seismic survey 海洋反射地震调查
marine refraction seismic survey 海洋折射地震调查
marine seismic acquisition 海上地震(数据)采集
marine seismic profiler 海洋地震剖面仪
marine seismic streamer 海洋地震漂浮电缆
marine seismic survey 海洋地震调查
marine wide-angle reflection seismic survey 海洋广角反射地震调查
mark 测标
mark for measuring velocity 测速标
Mars 火星
Mars cloud 火星云
Mars dayglow 火星日气晖
Mars dust storm 火星尘暴
Mars escape velocity 火星逃逸速度
Mars ionosphere 火星电离层
Mars polar cap 火星极盖
Marsquake 火星震
mask 蒙片
mask artwork 蒙绘
mass absorption coefficient 质量吸收系数
mass spectrograph 质谱仪
mass spectrometry 质谱学,质谱分析
master earthquake 主导地震
master event 主导事件
mathematical cartography 数学地图学

Matuyama epoch 松山期
maximum corona 极大日冕
maximum usable frequency (MUF) 最大可用频率
Maxwell distribution 麦克斯韦分布
Maxwell's equation 麦克斯韦方程
M discontinuity (Mohorovicic discontinuity) 莫霍(洛维契奇)界面, M界面
meadow 草甸
mean collision time 平均碰撞时间
mean earth ellipsoid 平均地球椭球
mean free path 平均自由程
mean motion 平均运动
mean pole 平极
mean pole of the epoch 历元平极
mean radius of curvature 平均曲率半径
mean sea level 平均海面
mean solar day 平太阳日
mean square error of angle observation 测角中误差
mean square error of side length 边长中误差
mean-time clock 平时钟
measured data 实测数据,测量数据
measurement 测量
measurement error 测量误差
measurement precision 测量精度
measuring bar 测杆
mechanical projection 机械投影
mechanical remanence 机械剩磁
mechanism of coronal heating 日冕加热机制
medium 介质
medium storm 中暴
Medvedev-Sponheuer-Karnik (intensity) scale 麦德维捷夫-施蓬霍伊尔-卡尔尼克(烈度)表, MSK表
meizoseismal area 极震区

mental map 心象地图,意境地图
Mercanton interval 默坎顿间段
Mercator chart 墨卡托海图
Mercator projection 墨卡托投影
Mercury 水星
merging efficiency 合并效率
meridian 子午圈
meridian passage 中天
meridian plane 子午面
meridional circulation 经向环流
meridional part 渐长纬度
meson producing layer 介子生成层
meson telescope 介子望远镜
mesopause 中间层顶
mesosphere 中间层
metadata 元数据
metallic spring gravimeter 金属弹簧重力仪
metamorphic water 变质水
metasphere 上气层
meteor bumper 流星防护屏
meteor dust 流星尘埃
meteor flare 流星爆发
meteoric water 雨水,大气水
meteorite crater 陨星坑
meteoritic astronomy 陨星天文学
meteoritics 陨星学
meteorological chart 气象图
meteorological representation error 气象代表误差
meteorology 气象学
meteor radar 流星雷达
meteor trail 流星余迹
methane 甲烷
methane production 甲烷生成
method by hour angle of Polaris 北极星任意时角法
method of deflection angle 偏角法
method of direction observation 方向观测法
method of laser alignment 激光准直法
method of tension wire alignment 引张线法
method of time determination by star transit 恒星中天测时法
method of time determination by Zinger star-pair 津格尔测时法,东西星等高测时法
metric camera 量测摄影机
Mg^+ reaction 镁离子反应
microcosm 微观宇宙
microdensitometer 测微密度计
microearthquake 微震
microfilm map 缩微地图
microgravimetry 微重力测量学
micrologging 微电极测井
micrometer 测微器
micrometer eyepiece 测微目镜
micropulsation 微脉动
microregionalization 小区划
microresistivity logging 微电极测井,微电阻率测井
microseism 微震,脉动
microseismic monitoring 微震监测
microseismic storm 脉动暴
microspot 微黑子
microvibrograph 微震仪
microwave radiation 微波辐射
microzonation 小区划
middle atmosphere 中层大气
middle magnetosphere 中磁层
middle tone 中性色调,灰色调
middleware 中间件
midnight aurora 子夜极光
midnight sector 子夜区段
mid-ocean ridge 洋中脊
migrating wave seismic facies 迁移波状地震相
migrating weather system 迁移气候系统
migration 偏移
migration imaging 偏移成像
migration velocity 偏移速度
migration velocity analysis 偏移速度分析
military chart 军用海图
military engineering survey 军事工程测量
military map 军用地图

Milne-Shaw seismograph 米尔恩-萧地震仪
Mimas 土卫一
mine 矿山
mineral deposits geometry 矿体几何学
mine survey 矿山测量
mine surveying 矿山测量学
mining area 矿区
mining engineering plan 采掘工程平面图
mining map 矿山测量图
mining subsidence observation 开采沉陷观测
mining theodolite 矿山经纬仪
mining yard plan 矿场平面图
minor angle method 小角度法
minor axis 短轴
MIP method (magnetic induced polarization method) 磁激发极化法
Miranda 天卫五
mirror altitude 镜点高度
mirror electron 镜点电子,反射电子
mirror field 镜点场
mirror latitude 镜点纬度
mirror point 镜像点
mirror reflection 镜反射
mirror reverse 反像
mirror separation 镜点间距
mise-a-la-masse method 充电法
mixed tidal harbor 混合潮港
mixer 混波器
MM (intensity) scale 修订的麦卡利(烈度)表,MM 表
mobile station 船台,移动台
mobility coefficient 迁移系数,迁移率
model chromosphere 色球模型
model photosphere 光球模型
moderate disturbance 中等扰动
moderate earthquake 中震
moderate geomagnetic storm 中等磁暴
mode-ray duality 振型-射线双重性
modified Mercalli (intensity) scale 修订的麦卡利(烈度)表,MM 表
mofette 碳酸气孔
Moho 莫霍(洛维契奇)界面,M 界面
Mohorovicic discontinuity (M discontinuity) 莫霍(洛维契奇)界面,M 界面
moire 龟纹
molecular absorption band 分子吸收带
molecular diffusion 分子扩散
molecular hydrogen 分子氢
molecular hydrogen chemistry 分子氢化学
molecular hydrogen spectrum 分子氢光谱
molecular nitrogen 分子氮
molecular oxygen 分子氧
molecular oxygen emission 分子氧发射
molecular oxygen spectrum 分子氧光谱
molecular weight 分子量
Molodensky formula 莫洛坚斯基公式
Molodensky theory 莫洛坚斯基理论
moment-density tensor 矩密度张量
moment magnitude 矩震级
moment tensor 矩张量
monitoring network 监测网
monitor record 监视记录
monitor station 监测台
monochromatic 单色的
Mono Lake excursion 莫诺湖漂移
monthly mean sea level 月平均海面
moonquake 月震
moon seismograph 月震仪
morphometric map 地貌形态示量图

motional induction 动生感应
mountain 山地
moveout 时差
moving source method 动源法
MSK (intensity) scale 麦德维捷夫-施蓬霍伊尔-卡尔尼克(烈度)表,MSK 表
MST radar 对流层、平流层、中层大气探测雷达,MST 雷达
mud 泥浆
mud logging 泥浆测井
mud spring 泥泉
MUF (maximum usable frequency) 最大可用频率
multichannel seismic instrument 多道地震仪
multidomain grain 多畴颗粒
multidomain thermal remanence 多畴热剩磁
multi layer organization 多层结构
multimedia map 多媒体地图
multiple coverage 多次覆盖
multiple earthquake 多重地震
multiple frequency amplitude-phase method 多频振幅相位法
multiple reflection 多次反射
multiple scattering 多次散射
multiplex 多路编排
multiplication constant 乘常数
multiplicity factor 倍增因子
multiplicity process 倍增过程
multiwave 多波
Mungo Lake excursion 蒙戈湖漂移
Munsell colour system 芒塞尔色系
Murray meteorite 莫里陨石
mute 消减(噪声)

Na (sodium) 钠
narrow band filter 窄带滤波器
national atlas 国家地图集
national leveling network 国家水准网
natural oscillation 固有振动
natural remanence 天然剩磁
natural remanent magnetization (NRM) 天然剩磁
navigation 导航
navigation obstruction 航行障碍物
navigation of aerial photography 航摄领航
navigation positioning 导航定位
navigation system 导航系统
N-axis N轴；B轴
neap rise 小潮升
near earthquake 近震
near field 近场
near-field seismology 近场地震学
near-infrared 近红外
near polar orbit 近极地轨道
near ultraviolet 近紫外
near UV excitation 近紫外激发
negative anomaly 负异常
negative correlation 负相关
negative image 阴像
neighbourhood method 邻元法
Neptune 海王星
Nereid 海卫二
network adjustment 网平差
neutral air 中性气体
neutral mass spectrometer 中性质谱仪
neutral surface 中性面
neutrino 中微子
neutron 中子
neutron activation logging 中子活化测井
neutron activation method 中子活化法
neutron-epithermal neutron logging 中子-超热中子测井
neutron-neutron logging 中子-中子测井
neutron star 中子星
neutron-thermal neutron logging 中子-热中子测井
neutron-γ logging 中子-γ测井
neutropause 中性层顶
neutrosphere 中性层
Newton's law 牛顿定律
night air-glow spectrum 夜气辉光谱
nightglow continuum 夜气辉连续谱
night hemisphere 背阳半球
nightside magnetosphere 夜侧磁层
night-sky radiation 夜天辐射
nitric acid 硝酸
nitric oxide 一氧化氮
nitric oxide ionization limit 一氧化氮电离限
nitric oxide production 一氧化氮生成
nitrogen atom 氮原子
nitrogen cycle 氮循环
nitrogen dioxide 二氧化氮
nitrogen dioxide photochemistry 二氧化氮光化学
nitrogen dioxide reaction 二氧化氮反应

N line N线
NMO correction (normal moveout correction) 动校正，正常时差校正
NMR (nuclear magnetic resonance) 核磁共振
NMR sounding 核磁共振测探
noctilucent cloud 夜光云
nocturnal ozone variation 夜间臭氧变化
nocturnal radiation 夜间辐射
nodal plane 节面
nom for selection 选取限额
nominal accuracy 标称精度
nominal scaling 名义量表
non-antiparallel magnetic fields 非逆平行磁场
non-coherent echo 非相干回波
non-deviative absorption 非偏移吸收
non-explosive source 非炸药震源
non-linear sweep 非线性扫描
non-linear wave 非线性波
non-ore anomaly 非矿异常
non-proton flare 非质子耀斑
non-thermal radiation 非热辐射
non-volcanic geothermal region 非火山地热区
non-zero plasma temperature 非零等离子体温度
normal component 法向分量
normal dispersion 正频散
normal distribution 正态分布
normal earthquake 正常（深度）地震
normal equation 正规方程
normal gravitation potential 正常引力位
normal gravity 正常重力
normal gravity field 正常重力场
normal gravity formula 正常重力公式
normal gravity potential 正常重力位
normal height 正常高
normalized total gravity gradient 归一化重力总梯度
normal level ellipsoid 正常水准椭球，水准椭球
normal mode 简正振型
normal moveout correction (NMO correction) 动校正，正常时差校正
normal polarity 正向极性
normal projection 正轴投影
normal section 法截面
normal shock wave 正激波
northern lights 北极光
northern polar cap 北极盖
north magnetic pole 磁北极
north-south asymmetry 南北不对称
nose structure 鼻形结构
nose whistler 鼻哨
notice to navigator 航行通告
NRM (natural remanent magnetization) 天然剩磁
nuclear astrophysics 核天体物理学
nuclear emulsion 核乳胶
nuclear geochemistry 核地球化学
nuclear magnetic resonance (NMR) 核磁共振
nuclear reaction 核反应
nucleating effect 致核效应
null vector 零向量，N轴
numerical solution of motion equation 运动方程数值解
Nunivak event 努尼瓦克事件
nutation 章动
nutation noise 章动噪音
nutation period 章动周期

Oberon 天卫四
oblique projection 斜轴投影
oblique shock 斜激波
oblique traces 斜截面法
obliquity change 倾斜度变化
observation 观测
observational error 观测误差
observation set 观测集
observatory 天文台
observed value 观测值
obtuse angle 钝角
occultation 掩
occultation of star 掩星
occurrence frequency 发生频次
ocean-bottom seismograph 海底地震仪
octupole 八极
offset 偏移;偏移距
offshore single-source and single-streamer seismic acquisition 单源单缆海上地震采集
oil exploration 石油勘探
oil reservoir 油藏
oil saturation 含油饱和度
Olduvai event 奥杜瓦伊事件
omnidirectional 全向的,非定向的
omnidirectional geophone pattern 全方位检波器组合
one-dimensional model 一维模式
one-hop propagation 一跳传播
opaque plasma 不透明等离子体
open configuration 开位形
open cube display 展开立体图
open drift orbit 开漂移轨道,不闭合漂移轨道
open photomultiplier 无窗光电倍增管

Ophelia 天卫七
opposite polarity 相反极性
optical air mass 大气光学质量
optical distance 光程
optical pump magnetometer 光泵磁力仪
optical thickness 光学厚度
optical window 光学窗
orbit 轨道
orbital angular momentum 轨道角动量
orbital decay 轨道衰变
orbital motion 轨道运动
orbital movement 轨道移动
orbital period 轨道周期
ordered perception 等级感
ordinal scaling 顺序量表
ordinary wave 寻常波
ordovician 奥陶系的
ordovician system 奥陶系
ore anomaly 矿异常
ore geochemical anomaly 成矿地球化学异常
organic aerosol 有机悬浮颗粒
organic geochemistry 有机地球化学
organic geochemistry method 有机地球化学法
organic scintillator 有机闪烁体
organ-pipe mode 风琴管振型
orientation 定向;方位
orientation of reference ellipsoid 参考椭球定位
orienteering map 定向运动地图
origin of coordinates 坐标原点
origin of longitude 经度起算点
origin time 发震时刻

orogenic geothermal belt 造山地热带
orthogonal function 正交函数
orthographic projection 正射投影
orthometric height 正高
oscillator strength 振子强度
osculating orbit 密切轨道
outbreak of flares 耀斑爆发
outer-core 外核
outer ionosphere 外电离层
outer magnetosphere 外磁层
outer radiation belt 外辐射带
outer space 外层空间
outgoing infrared radiation 射出红外辐射
outline map (for filling) 填充地图
out of phase 异相
outward flux 外流通量,向外通量
oval-shaped belt 卵形带
overall rate 总反应速率
overburden pressure 盖层压力
overcoring 套芯钻
Overhauser magnetometer 欧弗豪泽磁力仪,双重核共振磁力仪
overheating 过热
overlapping average 叠加平均
overlay 叠加
overlay interpretation 叠合解释
overpressure 超压
overprint 叠印
overtone normal mode 谐波简正振型
oxidation mechanism 氧化机制
oxygen geochemical barrier 氧地球化学障
ozone absorption 臭氧吸收
ozone amount 臭氧量
ozone concentration 臭氧浓度
ozone cutoff 臭氧截断
ozone distribution 臭氧分布
ozone formation 臭氧形成
ozone photolysis 臭氧光解
ozone reaction 臭氧反应
ozone spectrophotometer 臭氧分光光度计
ozonopause 臭氧层顶
ozonosphere 臭氧层

palaeogeomagnetic equator 古地磁赤道
palaeogeomagnetic intensity 古地磁强度
palaeogeothermics 古地热学
palaeolatitude 古纬度
palaeolongitude 古经度
palaeomagnetic direction 古地磁方向
palaeomagnetic field 古地磁场
palaeomagnetic pole 古地磁极
palaeomagnetism 古地磁(学)
panchromatic film 全色片
panchromatic infrared film 全色红外片
panorama camera 全景摄影机
panoramic distortion 全景畸变
panoramic photography 全景摄影
parallax 视差
parallel circle 平行圈
parallel conductivity 平行电导率
parameter adjustment 参数平差
parametric effect 参量效应
parent comet 母彗星
Parker model 帕克模式
PARM (partial ARM) 部分无滞剩磁
partial ARM (PARM) 部分无滞剩磁
partial eclipse 偏食
partial lunar eclipse 月偏食
partial solar eclipse 日偏食
partial thermoremanent magnetization (PTRM) 部分热剩磁
particle accelerator survey 粒子加速器测量
particle bombardment 粒子轰击
particle flux 粒子通量
particle precipitation zone 粒子沉降带
particle radiation 粒子辐射
particle scattering 粒子散射
particular map 特种地图
passive remote sensing 被动式遥感
passive source method 被动源(方)法
pass point 加密点
patch 斑块
patchy aurora 亮斑极光
Paterson reversal 帕特森反向
path of eclipse 食带
path of total eclipse 全食带
pattern 格局;模式
pattern change 格局变化
pattern recognition 模式识别
pattern recognition of remote sensing 遥感模式识别
P-axis (pressure axis) 压力轴, P轴
PCA (polar cap absorption) 极盖吸收
PCGIAP (Permanent Committee on GIS Infrastructure for Asia and the Pacific) 亚太区域地理信息系统基础设施常设委员会
P Code (Precise Code) 精码
PDE (Preliminary Determination of Epicentre) 初定震中
peak 峰值
peak acceleration 峰值加速度
peak altitude 峰值高度

peak concentration 峰值浓度
peak displacement 峰值位移
peak velocity 峰值速度
Pedersen conductivity 彼得森电导率
peel-coat film 撕膜片
pelagic survey 远海测量
penetrating radiation 穿透辐射,贯穿辐射
penetration depth 穿透深度,贯穿深度
penny-shaped crack 币形裂纹
penumbra cone 半影锥
percent frequency effect 百分频率效应
perceptual effect 感受效果
perceptual grouping 感知分组
periastron 近星点
pericenter 近心点
pericynthion 近月点
perigee 近地点
periodic comet 周期彗星
Permanent Committee on GIS Infrastructure for Asia and the Pacific (PCGIAP) 亚太区域地理信息系统基础设施常设委员会
permanent magnetization 永久磁化
permitted line 容许谱线
permitted transition 容许跃迁
perspective geochemical prospecting 预期地球化学勘探
perspective projection 透视投影
perspective traces 透视截面法
perturbation coefficient 扰动系数
phase (震)相
phase ambiguity 相位多值性
phase ambiguity resolution 相位模糊度解算
phase comparison 相位比较,比相
phase cycle 相位周
phase cycle value 相位周值
phase difference 相差
phase discrimination 震相辨别
phase drift 相位漂移
phase identification 震相识别
phase induced polarization method 相位激发极化法
phase of an eclipse 食相
phase of the moon 月相
phase scintillation 相位闪烁
phase stability 相位稳定性
phase transfer function (PTF) 相位传递函数
phase velocity 相速度
Phobos 火卫一
Phoebe 土卫九
photo base 像片基线
photochemical action 光化作用
photochemical effect 光化效应
photochemical reaction 光化学反应
photochemistry 光化学
photo coordinate system 像平面坐标系
photodetachment 光致脱离
photodissociation 光致离解
photoelectric current 光电流
photoelectric effect 光电效应
photoelectric photometer 光电光度计
photoelectron spectroscopy 光电子光谱学
photo-excitation 光致激发
photogrammetry 摄影测量
photographic zenith tube 照相天顶筒
photo interpretation 像片判读
photoionization 光致电离
photoionization cross section 光电离截面
photolysis 光解作用
photo mosaic 像片镶嵌
photo nadir point 像底点
photo orientation element 像片方位元素
photoplan 像片平面图
photo-recombination 光致复合
photo rectification 像片纠正
photo scale 像片比例尺

phototypesetter 照相排字机
physical geochemistry 物理地球化学
physical geodesy 物理大地测量学,大地重力学
physical map 自然地图
physical mechanism 物理机制
picto-line map 浮雕影像地图
picture format 像幅
piezo-magnetic effect 压磁效应
piezo-remanence 压剩磁
piezo-remanent magnetization (PRM) 压剩磁
pilot anchorage 引水锚地
pilot atlas 引航图集
pitch 航向倾角
pitch axis 俯仰轴
pixel 像素
plage flare 谱斑状耀斑
Planck law 普朗克定律
plane 平面图
plane curve location 平面曲线测设
plane of polarization 偏振面
plane polarization 平面偏振
plane-table 平板仪
plane-table survey 平板仪测量
plane-table traverse 平板仪导线
planetary aberration 行星光行差
planetary companion 类行星伴星
planetary configuration 行星动态
planetary cosmogony 行星演化学
planetary geodesy 行星(大地)测量学
planetary heat budget 行星热收支
planetary physics 行星物理学
planetary precession 行星岁差
planetary seismology 行星震学
planetary stream 行星流星雨
planetary wave 行星波
planetography 行星表面学
planet-wide geothermal belt 全球性地热带
planimeter 求积仪
planning map 规划地图

plantation 人工林
plasma ejection 等离子体抛射
plasma frame of reference 等离子体参考系
plasma frequency 等离子体频率
plasma instability 等离子体不稳定性
plasma mantle 等离子体幔
plasmapause 等离子体层顶
plasma physics 等离子体物理学
plasma resonance 等离子体共振
plasmasheet 等离子体片
plasmasphere 等离子体层
plastic scintillator 塑料闪烁体
plate 板块
plate collision 板块碰撞
plate copying 晒版
plate correction 层间改正
plate tectonics 板块(大地)构造学
plotter 绘图机
plotting file 绘图文件
plumb aligner 垂准仪,铅垂仪
plumb bob 垂球
plumb line 铅垂线
Pluto 冥王星
plutonic water 深成水
point coordinate positioning 极坐标定位,距离方位定位
point for shaft position 近井点
point mode 点方式
point position 点位
point symbol 点状符号
Poisson distribution 泊松分布
Poisson's equation 泊松方程
polar axis 极轴
polar cap absorption (PCA) 极盖吸收
polar cap aurora 极盖极光
polar cap glow 极盖气辉
polar coordinate 极坐标
polar flux tube 极区流量管
polarity bias 极性偏向
polarity chron 极性年代
polarity dating 极性年代测定
polarity epoch 极性期

polarity event 极性事件
polarity interval 极性间段
polarity sequence 极性序列
polarity subchron 极性亚代
polarity superchron 极性超代
polarity transition 极性过渡
polarization angle 偏振角
polarization effect 极化效应
polarization ellipse 偏振椭圆
polarization radiation 偏振辐射
polar magnetic disturbance 极区磁扰
polar motion 极移
polar pantograph 极坐标缩放仪
polar phase shift 极相漂移
polar plasma 极区等离子体
polar positioning method 极坐标定位方法,方位距离定位方法
polar ray 极射线
polar shower 极地簇射
polar wander 极移
polar-wander curve 极移曲线
polar-wander path (PWP) 极移路径
polar wind 极风
pole-dipole array 三极排列
pole of spreading 扩张极
pole of the ecliptic 黄极
poleward horizon 极向水平
political map 政治地图
poloidal 极型
poloidal oscillation 极型振荡
polyconic projection 多圆锥投影
polyfocal projection 多焦点投影
polygon 多边形
polygonal map 多边形地图
polygon structure 多边形结构
polynomial 多项式
polytropic index 多层球指数
population map 人口地图
porosity 孔隙度
Porro-Koppe principle 波罗-科普原理
Portia 天卫十二
positioning 定位
positioning diagram method 定位统计图表法
positioning precision 定位精度
positioning system 定位系统
positioning technology 定位技术
positive anomaly 正异常
positive image 阳像,正像
positive ion 阳离子,正离子
positive temperature effect 正温效应
positron 阳电子,正电子
post-depositional DRM 沉积后碎屑剩磁
post glacial rebound 冰后回弹
post-glacial time 冰后期
post-seismic 震后的
post stack migration 叠后偏移
posture map 态势地图
potential 电位
potential electrode 测量电极
potential energy 势能
potential gradient 电位梯度
potential of centrifugal force 离心力位
Potsdam gravimetric system 波茨坦重力系统
power density spectrum 功率密度谱
powered flight 主动段飞行
power-law creep 幂次律蠕变
power spectrum 功率谱
PRARE (Precise Range and Rangerate Equipment) 普拉烈系统
Pratt-Hayford isostasy 普拉特-海福德均衡
preceding spot 前导黑子
precipitating electron 沉降电子
precipitation 降水
precipitation flux 沉降通量
precise alignment 精密准直
Precise Code (P Code) 精码
precise engineering control network 精密工程控制网
precise engineering survey 精密工程测量

precise ephemeris 精密星历
precise level 精密水准仪
precise leveling 精密水准测量
precise mechanism installation survey 精密机械安装测量
precise plumbing 精密垂准
Precise Range and Rangerate Equipment (PRARE) 普拉烈系统
precise ranging 精密测距
precise survey at seismic station 地震台精密测量
precise traversing 精密导线测量
precision 精度
precision estimation 精密估计
precision stereoplotter 精密立体测图仪
precursor 前兆
precursor time 前兆时间
precursory 前兆的
predawn enhancement 黎明前增强
prediction 预报
predictive deconvolution 预测反褶积
predictive indicator 预报因子
predissociation 预离解
Preliminary Determination of Epicentre (PDE) 初定震中
Preliminary Reference Earth Model (PREM) 初始参考地球模型
PREM (Preliminary Reference Earth Model) 初始参考地球模型
pre-press proof 预打样图
preprinted symbol 预制符号
pre-seismic 震前的
presensitized plate 预制感光版,PS版
Press-Ewing seismograph 普雷斯-尤因地震仪
pressure axis (P-axis) 压力轴,P轴
pressure gauge 压力验潮仪
pressure gradiant 压力梯度

pressure scale height 压力标高
prestack migration 叠前偏移
primary cosmic radiation 原宇宙辐射
primary geochemical anomaly 原生地球化学异常
primary graphic element 基本图形元素
primary magnetization 原生磁化(强度)
primary remanent magnetization 原生剩磁
primary wave 初至波,P波
prime meridian 本初子午线
prime vertical 卯酉圈
prime vertical plane 卯酉面
primitive atmosphere 原始大气圈,上古代大气
primordial atmosphere 初生大气
principal distance of photo 像片主距
principal line 像主纵线
principal point of photograph 像主点
principle of equivalence 等价性原理
principle of geochemical distribution 地球化学的分布法则
principle of geometric reverse 几何反转原理
principle of least squares 最小二乘法原理
printer lens 照相制版镜头
printing down 晒版
printing plate 印刷版
PRM (piezo-remanent magnetization) 压剩磁
probability distribution 概率分布
probable error 概率误差
processional motion 进动
process lens 照相制版镜头
production seismic (油田)开发地震
product standard of digital maps 数字地图产品标准
profile 剖面
profile map 剖面图

prognostic map 预报地图
prograde 正转
progressive wave 前进波
prohibited area 禁航区
projection 投影
projection interval 渐长区间
projection transformation 投影变换
projection with two standard parallels 双标准纬线投影
Prometheus 土卫十六
prominence of the sunspot type 黑子型日斑,黑子日珥
prominence spectroscope 日珥分光镜
proofing 打样
propagation 传播
propagation delay 传播延迟
propogator matrix 传播矩阵
proportional counter 正比计数器
proportional error 比例误差
prospecting 找矿
prospecting baseline 勘探基线
prospecting line profile map 勘探线剖面图
prospecting line survey 勘探线测量
prospecting network layout 勘探网测设
proton 质子
proton aurora 质子极光
proton energy flux 质子能量流通量
proton flare 质子耀斑
proton magnetometer 质子磁强计
protonosphere 质子层
proton-precession magnetometer 质子旋进磁力仪
proton reaction 质子反应
pseudo-acceleration response spectrum 伪加速度反应谱
pseudo-aftershock 假余震
pseudogravity anomaly 磁源重力异常
pseudo-isoline map 伪等值线地图
pseudosection map 拟断面图
pseudo-trapped particle 假捕获粒子
pseudo-trapping region 伪捕区
pseudo-velocity response spectrum 伪速度反应谱
PTF (phase transfer function) 相位传递函数
PTRM (partial thermoremanent magnetization) 部分热剩磁
Puck 天卫十五
pulsating aurora 脉动极光
pulsation 脉动
pulsation theory 脉动理论,脉动学说
pulse frequency 脉冲频率
pulse height spectrum 脉高谱,脉冲高度谱
pulse recurrence frequency 脉冲重复频率
pulse shape discrimination 脉形甄别
pulse shaping 脉冲整形
pumping station 泵站
pure gravity anomaly 纯重力异常
pure sound wave 纯声波
purple soil 紫色土
putizze 硫化氢气孔
PWP (polar-wander path) 极移路径
pyramid 金字塔
pyranometer 辐射强度表

quadrant 象限仪
quadrupole 四极子
qualitative analysis 定性分析
qualitative perception 质量感
quality base method 质底法
quality of aerophotography 航摄质量
quantitative analysis 定量分析
quantitative perception 数量感
quantity base method 量底法
quantum yield 量子产额
quarantine anchorage 检疫锚地
quasi-longitudinal propagation 准纵传播
quasi-stable adjustment 拟稳平差
quasi-stellar object 类星体
quasi-transverse propagation 准横传播
quaternary 第四纪
quenching agent 猝灭剂
quenching reaction 猝灭反应
query 查询
quiescence spectrum 宁静光谱
quiescent primary radiation 宁静初级辐射
quiescent prominence 宁静日珥
quiet condition 宁静条件
quiet solar wind 宁静太阳风

radar altimeter 雷达测高仪
radar aurora 雷达极光
radar beacon 雷达信标
radar overlay 雷达覆盖区
radar ramark 雷达指向标
radar responder 雷达应答器
radial component 径向分量
radial diffusion 径向扩散
radial distortion 径向畸变
radial motion 径向运动
radial oscillation 径向振荡
radiant flux 辐射通量
radiant intensity 辐射强度
radiant power 辐射功率
radiation belt 辐射带
radiation budget 辐射收支
radiation cooling 辐射冷却
radiation coupling 辐射耦合
radiation damping 辐射阻尼
radiation effect 辐射效应
radiation medicine 放射医学
radiation pattern 辐射图型
radiation temperature 辐射温度
radiative attachment 辐射附着
radiative lifetime 辐射寿命
radiative pressure 辐射压
radiative recombination 辐射复合
radiative transfer 辐射传输
radioactive decay 放射性衰变
radioactive fallout 放射性沉降物
radioactive tracer logging 放射性示踪测井
radioactivity logging 放射性测井
radioactivity prospecting 放射性勘探
radioactivity survey 放射性调查

radio aurora 射电极光
radio brightness 射电亮度
radiocarbon 放射性碳
radioeclipse 射电食
radio emission 射电辐射
radio flux 射电通量
radio-isophote 射电等强线
radioisotope logging 同位素测井
radiometer 辐射计,放射计
radio-phase method 无线电相位法
radio plage 射电谱斑
radio spectrum 射电谱
radio star 射电星
radio telemetry seismic data acquisition 无线电遥测地震(数据)采集
radio window 射电窗
radius of curvature 曲率半径
radius of curvature in meridian 子午圈曲率半径
radius of curvature in prime vertical 卯酉圈曲率半径
radius vector 径矢,径向矢量
radon survey 氡气测量
rainfall 降雨
rake 滑动角
Raman effect 拉曼效应
random distribution 随机分布
random error 随机误差
random noise 随机噪声
range-energy curve 射程能量曲线
rangefinder 测距仪
range hole 测距盲区
range index 变幅指数
range-only radar 测距雷达

range positioning system 测距定位系统，圆-圆定位系统
range-range positioning 圆-圆定位，距离-距离定位
ranging 测距
rapid diffusion 快速扩散
rapidly moving aurora 速移极光
rarefaction region 稀疏区
rarefied plasma 稀薄等离子体
raster 栅格，光栅
raster data 栅格数据
raster plotting 栅格绘图
rate coefficient 速率系数
rate equation 速率方程
ratio enhancement 比值增强
ratio scaling 比例量表
ratio transformation 比值变换
ray 射线
rayed arc 射线极光弧
ray equation 射线方程
Rayleigh limit 瑞利极限
Rayleigh wave 瑞利波，R波
ray method 射线法
ray parameter 射线参数
ray structure 射线状结构
ray tracing 射线追踪
reaction chain 反应链
reaction rate coefficient 反应速率系数
reading accuracy of sounder 测深仪读数精度
real-aperture radar 真实孔径雷达
real current system 实际电流体系
real dipole 真偶极子
real geomagnetic field 实际地磁场
real time correlation 实时相关
real time data 实时数据
real time telemetry 实时遥测
recapture 再俘获
receiver statics 接收点静校正
receiving 接收
receiving centre 接收中心
reception diode 接收二极管

recombination coefficient 复合系数
recombination radiation 复合辐射
recombination rate 复合率
reconnection 重联
reconnection rate 重联速率
recording paper of sounder 测深仪记录纸
recovery phase 恢复相
rectangular map subdivision 矩形分幅
rectification 纠正
rectifier 纠正仪，整流器
recurrence frequency 重现频率
recurrent period 重现周期，循环周期
red arc 红弧
red aurora 红极光
red shift 红移
red spot 红斑
reduced heat flow 折合热流量
reduced latitude 归化纬度
reduced pendulum length 折合摆长
reduced to the magnetic pole 磁极归化
reduced travel time 折合走时
reducing colour printing 减色印刷
reduction geochemical barrier 还原地球化学障
reef 暗礁
reentry mode 再入式
reentry trajectory 再入轨道
reentry vehicle 再入飞行器
reference altitude 参考高度
reference effect 参照效应
reference ellipsoid 参考椭球
reference value 参考值
reflected frequency 反射频率
reflecting effect 反射效应
reflection 反射
reflection coefficient 反射系数
reflection matrix 反射矩阵
reflection seismology 反射地震学
reflection wave 反射波

reflectivity method 反射率法
refraction 折射,折光
refraction correlation method 折射波对比法
refractive index 折射率
regional anomaly 区域异常
regional atlas 区域地图集
regional earthquake 区域地震
regional geochemical anomaly 区域地球化学异常
regional geochemical background 区域地球化学背景
regional geochemical differentiation 区域地球化学分异
regional geochemical prospecting 区域地球化学勘探
regional geochemistry 区域地球化学
regional geological map 区域地质图
regional geological survey 区域地质调查
regionalization map 区划地图
regional seismic stratigraphy 区域地震地层学
register mark 规矩线
registration 对准,配准
regression analysis 回归分析
regression equation 回归方程
regression estimation 回归估计
relative abundance 相对丰度
relative amplitude preserve 相对振幅保持
relative error 相对误差
relative flying height 相对航高
relative gravity measurement 相对重力测量
relative orientation 相对定向
relative ozone concentration 相对臭氧浓度
relative positioning 相对定位
relativistic astrophysics 相对论性天体物理学
relativistic correction 相对论改正
relativistic event 相对论性天体事件
relativistic mass 相对论质量
relaxation source 松弛源
relaxation time 弛豫时间
relief map 立体地图
remagnetization 再磁化
remagnetization circle 再磁化圆(弧)
remanence 剩余磁化
remanent magnetization 剩余磁化(强度)
remote resonance 远谐振,远共振
remote sensing 遥感
remote sensing data acquisition 遥感数据获取
remote sensing mapping 遥感制图
remote sensing platform 遥感平台
remote sensing sounding 遥感测深
replicative symbol 象形符号
resection 后方交会
reservoir 储集层,水库
reservoir-induced earthquake 水库地震
reservoir seismic stratigraphy 储层地震地层学
reservoir storage survey 库容测量
reservoir-triggered seismicity 水库触发地震
residual flux density 剩磁通量密度
residual gravity anomaly 剩余重力异常
residual range 剩余射程
resistive element 电阻性元件
resistivity 电阻率
resistivity logging 电阻率测井
resistivity method 电阻率法
resistivity profiling 电阻率剖面法
resolution 分辨率
resolution acuity 视觉分辨敏锐度

resonance line 共振线
resonance probe 共振探针
resonance radiation 共振辐射
resonant scattering 共振散射
respiration rate 呼吸速率
rest frame 静止参考系
rest mass 静止质量
restricted area 限航区
retouching 修版
retrieval by header 定性检索
retrieval by window 定位检索, 开窗检索
retrograde 倒转
return period 重现周期
Reunion event 留尼旺事件
reversal points method 逆转点法
reversal test 倒转检验
reverse branch 回转波
reversed polarity 反向极性
reversible process 可逆过程
Reynolds' number 雷诺数
Rhea 土卫五
rheological intrusion 流变性侵入体
Richter magnitude 里氏震级
ridge-type earthquake 洋脊型地震
right ascension 赤经
right-handed orthogonal system 右手正交系
right-reading 正像
rigid boom system 硬架系统
rigid frame system 硬架系统
rigidity spectrum 刚度谱
ring current 环电流
ringing 鸣震
riometer 宇宙噪声吸收仪
river basin 河流域
river channel 河道
river chart 江河图
river-crossing leveling 跨河水准测量
river improvement survey 河道整治测量
river network 河网
river survey 江河测量

robust estimation 抗差估计, 稳健估计
Roche limit 洛希极限
rock 礁石
rock burst 岩爆
rocket balloon instrument 火箭气球装置
rocket-borne mass spectrometer 箭载质谱计
rocket-launching site 火箭发射场
rocket meteorograph 箭载气象仪
rocket propellant 火箭推进器, 火箭推进剂
rock magnetism 岩石磁性
rocky desertification 石漠化
rod 标尺
Roelofs solar prism 鲁洛夫斯太阳棱镜
Roentgen-equivalent-man 人体生物伦琴当量
roll axis 滚动轴
root-mean-square error 均方根误差
root of mountain 山根
Rosalind 天卫十三
Rossby wave 罗斯比波
Rossi-Forel (intensity) scale 罗西-福勒烈度表, RF 表
Rossi-Forel scale 罗西-福勒表, RF 表
rotating frame of reference 转动参考系
rotational discontinuity 转动间断
rotational line 转动谱线
rotational quantum number 转动量子数
rotational remanence 旋转剩磁
rotational remanent magnetization (RRM) 旋转剩磁
rotational wave 旋转波
rotation angle 旋转角
rotation parameter 旋转参数
roughness 糙率
round-off error 舍入误差
RRM (rotational remanent magnetization) 旋转剩磁

Ru (ruthenium) 钌
rubidium magnetometer 铷蒸气磁强计
ruling 网线
run error 行差
runoff 径流
runoff yield 产流
rupture 破裂
rupture front 破裂前沿
rupture length 破裂长度
rupture process 破裂过程
rupture propagation 破裂传播
rural planning survey 乡村规划测量
ruthenium (Ru) 钌

sailing chart 航行图
sailing direction (SD) 航路指南
saline 盐渍
saline-alkaline 盐碱
salinity 盐分,盐度
salinization 盐渍化
salt content 含盐量
Salyut Space Station 礼炮号航天站
sampling 采样
sampling interval 采样间隔
sand 砂
sandstone 砂岩
sandstone reservoir 砂岩储层
sandwich construction 夹层结构
sandy 沙地的
SAR (Synthetic Aperture Radar) 合成孔径雷达
SAR image 合成孔径雷达图像
satellite altimetry 卫星测高
satellite-borne sensor 星载遥感器
satellite geodesy 卫星大地测量学
satellite infrared data 卫星红外资料
satellite lifetime prediction 卫星寿命预报
satellite meteorology 卫星气象学
satellite orbit 卫星轨道
satellite scintillation 卫星闪烁
satellite system 卫星系统
saturation 饱和度
Saturn 土星
Saturnian satellite 土卫
Saturn's ring 土星光环
scalar intensity 标量强度

scalar magnetometer 标量磁强计
scalar multiplication 标量乘法,标乘
scalar quantity 标量
scale 比例尺
scale height 标高
scale parameter 尺度参数
scaling law 尺度定律
scaling of model 模型缩放
scan-digitizing 扫描数字化
scanning microwave spectrometer 扫描微波光谱仪
scanning spectrometer 扫描分光计
scattered γ-ray logging 散射γ测井
scattering 散射
scattering angle 散射角
scattering coefficient 散射系数
scattering probability 散射几率
Scheimpflug condition 交线条件,向甫鲁条件,恰普斯基条件
Schlumberger array 施伦伯格排列
Schlumberger electrode array 施伦伯格电极排列
Schmidt camera 施密特照相机
school map 教学地图
Schreiber method in all combinations 施赖伯全组合测角法
Schumann resonance 舒曼共振
Schumann UV 舒曼紫外
scintillator 闪烁体
scour 冲刷
scratcher electrode logging 滑动接触法测井

screen 网屏
screening height 屏蔽高度
screen map 屏幕地图
screw dislocation 螺型位错
scriber 刻图仪
scribing 刻绘
SD (sailing direction) 航路指南
SDI (spatial data infrastructure) 空间数据基础设施
sea floor spreading 海底扩张
sea gravimeter 海洋重力仪
seaquake 海震
sea shock 海震
seasonal correction of mean sea level 平均海面归算
seasonal effect 季节效应
secondary electron 次级电子
secondary geochemical anomaly 次生地球化学异常
secondary ion 次级离子
secondary magnetization 次生磁化(强度)
secondary radiation 次级辐射
secondary remanent magnetization 次生剩磁
secondary wave 续至波,S波
second law of thermodynamics 热力学第二定律
section 断面
section map 断面图
sector boundary 扇形边界
sector structure 扇形结构
secular change 长期变化
secular variation 长期变化
sediment 泥沙,淤积物
sedimentary geochemistry 沉积物地球化学
sedimentation 淤积
sediment-carrying 挟沙的
sediment concentration 含沙量
sediment movement 泥沙运动
sediment yield 产沙(量)
seepage 渗流
segmentation 分割
seismic absorption band 地震吸收带

seismic acceleration 地震加速度
seismic activity 地震活动性
seismically active belt 地震活动带
seismically active zone 地震活动区
seismic belt 地震带
seismic body wave 地震体波
seismic channel 地震道
seismic coefficient method 地震系数法
seismic cycle 地震轮回
seismic data preprocessing 地震数据预处理
seismic dislocation 地震位错
seismic earth pressure 地震动土压力
seismic efficiency 地震效率
seismic energy 地震能量
seismic exploration 地震勘探
seismic facies 地震相(勘探)
seismic facies analysis 地震相分析
seismic facies map 地震相图
seismic facies unit 地震相单元
seismic fault 地震断层
seismic gap 地震空区
seismic geology 地震地质学
seismic geophysical survey 地物探法
seismic ground motion 地震动
seismic hazard 震灾
seismic horizon 地震层位
seismic intensity 地震烈度
seismic inversion 地震反演
seismicity 地震活动性
seismicity pattern 地震活动性图像
seismic marker horizon 地震标准层
seismic mass 地震量
seismic model 地震模型
seismic moment 地震矩
seismic network 地震台网
seismic parameter 地震参数
seismic phase 震相

seismic prospecting 地震勘探
seismic ray 地震射线
seismic record 地震记录
seismic recording instrument 地震(记录)仪
seismic reflection 地震反射
seismic reflection method 地震反射法
seismic refraction method 地震折射法
seismic regime 震情
seismic regionalization 地震区划
Seismic Research Observatory (SRO) 地震研究观测台
seismic reservoir study 地震储层研究
seismic response 地震响应
seismic risk 地震危险性
seismic risk analysis 地震危险性分析
seismic risk evaluation 地震危险性评定
seismic rupture 地震破裂
seismic sea wave 地震海啸
seismic section 地震剖面
seismic seiche 湖震
seismic sequence 地震序列
seismic signal 地震信号
seismic sounding 地震测深
seismic source 震源
seismic source dynamics 震源动力学
seismic source kinematics 震源运动学
seismic source parameter 震源参数
seismic station 地震台,地震站
seismic streamer 地震(勘探)等浮电缆
seismic structural map 地震构造图
seismic structural wall 抗震墙
seismic surface wave 地震面波
seismic surveillance 地震监测
seismic survey 地震调查
seismic survey vessel 地震勘探船
seismic technology 地震技术
seismic tectonics 地震构造学
seismic-tectonic zone 地震构造带
seismic tomography 层析地震成像
seismic trigger 地震触发器
seismic velocity 地震速度
seismic wave 地震波
seismic-wave dispersion 地震波频散
seismic wavelet 地震子波
seismic zone 地震区
seismic zoning 地震区划
seismogenic zone 孕震区
seismogeology 地震地质学
seismogram 地震图
seismograph 地震仪
seismological table 走时表
seismology 地震学
seismology model 地震模型(学)
seismometer 地震计
seismometry 测震学
seismoscope 验震器
seismosociology 地震社会学
seismotectonic province 地震构造区
seismotectonics 地震构造学
selection effect 选择效应
selective γ-γ logging 选择 γ-γ 测井
selenodesy 月面测量(学)
selenography 月面学
selenology 月球学
self-potential logging (SP logging) 自然电位测井
self-potential method 自然电位法
self-reversal 自反向
semi-annual effect 半年效应
semiconductor laser 半导体激光器
semidiurnal tidal component 半日潮汐分量
semidiurnal tidal harbour 半日潮港

semidiurnal variation 半日变化
semi-meridian 半子午线
semi-tropical zone 副热带
separate energy channel 分立能通道
sequential adjustment 序贯平差
series maps 系列地图
setting accuracy 安平精度
setting-out of cross line through shaft centre 井筒十字中线标定
settlement observation 沉降观测
settling 沉降
severe drought 大旱
sextant 六分仪
SFAP (small format aerial photography) 小像幅航空摄影
SFD (sudden frequency deviation) (短波)频率急偏
sferics 天电学
sferics fix 天电定位
sferics observation 天电观测
sferics receiver 天电接收器
shade 深色调
shadow transit 卫影凌行星
shadow zone 影区
shaft bottom plan 井底车场平面图
shaft orientation survey 立井定向测量
shaft prospecting engineering survey 井深工程测量
shallow-focus earthquake 浅(源地)震
shallow water seismic 浅海地震勘探
sharp boundary 锐边界
shear coupled PL waves (剪切耦合)PL波
shear dislocation 剪切位错
shear effect 剪切效应
shear instability 剪切不稳定性
shear melting 剪切熔融
shear wave 剪切波
sheet designation 图幅编号
sheet drape seismic facies 席状披盖地震相
sheet index 图幅接合表
sheet number 图幅编号
sheet seismic facies 席状地震相
shell splitting 壳分裂
Shida's number 志田数
shielding effect 屏蔽效应
shipboard gravimeter 船载重力仪
shoal 浅滩
shock absorption 消振
shock resistant 抗震的
shock size 地震大小
shock spectrum 激波谱
shock strength 激波强度
shock tunnel 激波风洞
shock wave 激波
shooting method 发射法
shoot statics 炮点静校正
short circuits 短路
short-period comet 短周期彗星
short-range positioning system 近程定位系统
short-term fluctuation 短期涨落
short-term response 短期响应
short wave fade-out (SWF) 短波衰退
short wave radiation 短波辐射
shot-geophone distance 炮检距
shower meteor 属群流星
shutter 快门
SID (sudden ionospheric disturbance) 突发电离层骚扰
side intersection 侧方交会
side-looking radar 侧视雷达
side overlap 旁向重叠
sidereal clock 恒时钟,恒星钟
sidereal day 恒星日
sidereal time 恒星时
sidereal year 恒星年
side scan sonar 侧扫声呐
side slope 边坡
side swipe 侧击波
Sidutjall event 西杜杰尔事件
sighting centring 照准点归心
sighting line method 瞄直法
sighting point 照准点

signal lamp 回光灯,标志灯
signal pole 信号杆
signal-to-noise 信噪比
significance level 显著水平
silent earthquake 寂静地震
silk-screen printing 丝网印刷
simple harmonic wave 简谐波
simultaneous contrast 视场对比
simultaneous observation 同步观测
sine curve 正弦曲线
single domain particle 单畴颗粒
single point 单点
single-stage rocket 单级火箭
single well 单井
site intensity 场地烈度
site remanence 原地剩磁
skip distance 跳距
sky background 天空背景
sky patrol 巡天观测
slant stack 倾斜叠加
slip 滑动
slip function 滑动函数
slip vector 滑动向量
slope aspect 坡向
slope gradient 坡度
slope line 示坡线
slope surface 坡面
slope theodolite 坡面经纬仪
slow drift wave 慢漂移波
slowness 慢度
slowness method 慢度法
slump seismic facies 滑塌地震相
slump test 坍塌检验
SM (surveying and mapping) 测绘学
small format aerial photography (SFAP) 小像幅航空摄影
small watershed 小流域
SMART 1 (Strong-Motion Array in Taiwan Number 1) 台湾强地动一号台阵
smoked paper record 熏烟纸记录图
smoke trails 烟迹
sodium (Na) 钠

sodium atomic emission 钠原子发射
sodium cloud 钠云
sodium layer 钠层
SOFAR channel (sound fixing and ranging channel) 声学定位测距声道,SOFAR 声道
soft-component 软成分
softer spectrum 较软能谱
soft particle 软粒子
soft radiation 软辐射
soft X-ray flux 软 X 线通量
soil conservation 水土保持
soil environment 土壤环境
soil erosion 土壤侵蚀
soil fertility 土壤肥力
soil layer 土层
soil microorganism 土壤微生物
soil nutrient 土壤养分
soil physical and chemical property 土壤理化性质
soil property 土壤性质
soil resource 土壤资源
soil rhizosphere 根圈,根际
soil salinity 土壤盐渍度,土壤盐分
soil salinization 土壤盐渍化
soil sample 土壤样品,土样
soil section 土壤剖面
soil surface 土壤表层
soil system classification 土壤系统分类
soil texture 土壤质地
soil type 土壤类型
solar activity 太阳活动
solar activity effect 太阳活动效应
solar activity event 太阳活动事件
solar activity impulse 太阳活动脉冲
solar activity minimum 太阳活动最小值
solar apex 太阳向点
solar astronomer 太阳天文学家
solar battery 太阳能电池

solar constant 太阳常数
solar control 太阳控制
solar corona 日冕
solar corpuscular emission 太阳微粒发射
solar cosmic ray 太阳宇宙线
solar cycle fluctuation 太阳活动周涨落
solar daily variation 太阳日变化
solar disk 日面
solar disturbance 太阳扰动
solar eclipse effect 日食效应
solar ecliptic coordinate 太阳黄道坐标
solar electromagnetic radiation 太阳电磁辐射
solar electron event 太阳电子事件
solar emission spectrum 太阳发射谱
solar energy spectrum 太阳能谱
solar equatorial plane 太阳赤道面
solar event 太阳事件
solar flare 太阳耀斑
solar flare activity 耀斑活动
solar flare cosmic radiation 耀斑辐射
solar flare cosmic ray 耀斑宇宙线
solar flare crochet 耀斑钩扰
solar flare disturbance 耀斑扰动
solar flare effect 耀斑效应
solar flare proton event 耀斑质子事件
solar gravitational field 太阳引力场
solar index 太阳指数
solar ionizing radiation 太阳离辐射
solar latitude 日面纬度
solar longitude 日面经度
solar magnetograph 太阳磁象仪
solar magnetospheric coordinate 太阳磁层坐标
solar maximum 太阳极大值
solar maximum year 太阳极大年
solar-meteorological relationship 太阳气象关系
solar neutrino unit 太阳中微子单位
solar occultation measurement 掩日测量
solar parameter 太阳参数
solar particle event 太阳粒子事件
solar phase angle 太阳相角
solar photon flux 太阳光子通量
solar photosphere 太阳光球
solar physicist 太阳物理学家
solar plasma 太阳等离子体
solar prominence 日珥
solar proton 太阳质子
solar proton event 太阳质子事件
solar radiation 太阳辐射
solar radiation flux 太阳辐射通量
solar radio emission 太阳射电发射
solar radio index 太阳射电指数
solar radio wave 太阳射电波
solar side 向阳侧
solar spectrograph 太阳光谱仪
solar spectrum 太阳光谱
solar system 太阳系
solar-terrestrial 日地的
solar-terrestrial physics 日地物理学
solar-terrestrial space 日地空间
solar-terrestrial storm 日地暴
solar tide 太阳潮
solar tide potential 太阳潮汐位能
solar ultraviolet radiation 太阳紫外辐射
solar-weather field 太阳-天气场
solar wind 太阳风
solar wind flow 太阳风流
solar wind magnetic field 太阳风磁场
solar wind particle 太阳风粒子
solar X-ray 太阳 X 射线

solar year 太阳年
solfatara 硫质气孔
solid earth geophysics 固体地球物理学
solitary charged particle 单个带电粒子
sonic logging 声波测井
sonogram 频时图
sound-emitting fireball 发声火流星
sound fixing and ranging channel (SOFAR channel) 声学定位测距声道,SOFAR 声道
sounding 测深(法)
sounding balloon 探空气球
sounding of induced polarization 激发极化测深
sounding pole 测深杆
sounding rocket 探空火箭
sounding system 探空系统
source array 组合源
source time function 震源时间函数
southbound node 降交点
south dipole pole 南偶极子极
southern polar cap 南极盖
south magnetic pole 磁南极
southseeking pole 指南极
southward field 南向磁场
southward value 南向值
Soyuz Spacecraft 联盟号宇宙飞船
SPA (sudden phase anomaly) 突发相位异常
space-based system 空基系统
space charge sheath 空间电荷鞘
space charge wave 空间电荷波
space chemistry 空间化学
spacecraft magnetic field 太空飞行体磁场
space density 空间密度
space electricity 空间电学
space geodesy 空间大地测量学
space intersection 空间前方交会
Spacelab 空间试验室
space magnetism 空间磁学

space optics 空间光学
space photogrammetry 航天摄影测量,太空摄影测量
space photography 航天摄影
space physics 空间物理学
space remote sensing 航天遥感
space resection 空间后方交会
space shuttle 航天飞机
spark spectrum 电花光谱
spatial analysis 空间分析
spatial data 空间数据
spatial database 空间数据库
spatial database management system 空间数据管理系统
spatial data infrastructure (SDI) 空间数据基础设施
spatial data transfer 空间数据转换
spatial pattern 空间格局
special use map 专用地图
specification of surveys 测量规范
specific gravity 比重
spectral albedo 光谱反照率,分光反照率
spectral channel 光谱通道,分光通道
spectral function 谱函数
spectral index 谱指数
spectral induced polarization method 频谱激发极化法
spectral line intensity 谱线强度
spectral logging 能谱测井
spectral region 光谱区
spectral series 谱线系
spectral term 光谱项
spectrohelioscope 太阳光谱观测镜
spectrometer 波谱测定仪
spectrophotometry 分光光度学
spectroscopy 光谱学
spectrum character curve 波谱特征曲线
spectrum cluster 波谱集群
spectrum feature space 波谱特征空间
spectrum of turbulence 湍流谱

spectrum response curve 波谱响应曲线
specular reflection 镜反射
spherical angle 球面角
spherical coordinate 球面坐标
spherical divergence compensation 球面发散补偿
spherical harmonic analysis 球谐分析
spherical harmonic expansion 球谐展开
spherical harmonic term 球谐项
spheroidal 球型
spheroidal oscillation 球型振荡
spike deconvolution 脉冲反褶积
spillway 泄洪道
spin axis 自旋轴
spinner magnetometer 旋转磁强计
spinning orbit 自旋轨道
spin rate 自旋率
spin stabilization 自旋稳定
spiral angle 螺旋角
spiral arm 旋臂
spiral curve location 缓和曲线测设
spiral orbit 螺旋轨道
spiral structure 螺旋状结构,旋涡结构
splitting 分裂
splitting parameter 分裂参数
SP logging (self-potential logging) 自然电位测井
spontaneous fault rupture 自发断层破裂
spontaneous emission 自发发射
spontaneous radiation 自发辐射
spontaneous rupture 自发破裂
sporadic E 散见E层
sporadic geomagnetic storm 偶发地磁暴
sporadic meteor 偶现流星
spot magnetic field 黑子磁场
spot maximum 黑子极大值
spot minimum 黑子极小值
spray prominence 喷射日珥

spread （检波器）排列
spread F 扩展F
spreading rate 扩张（速）率
spring maximum 春季极大值
square 二乘
SQUID magnetometer (superconductive magnetometer) 超导磁力仪
SRO (Seismic Research Observatory) 地震研究观测台
stable auroral red arc 稳压极光红弧
stable isotope geochemistry 稳定同位素地球化学
stable oscillation 稳定振动
stable trapping 稳定捕捉
stacked profiles map 叠加剖面图
stacking 叠加
stacking velocity 叠加速度
staff 标尺
stagnation pressure 驻点压力
standard deviation 标准偏差
standard error 标准误差
standard field of length 长度标准检定场
standard frequency 标准频率
standard-height analysis 标准高度分析
standard meridian 标准子午线
standard of surveying and mapping 测绘标准
standard parallel 标准纬线
standard station 基准台
standing shock front 驻激波锋面
standing wave 驻波
Starfish radiation belt 星鱼辐射带
starting phase 起始相
state of the magnetosphere 磁层状态
statical mechanical magnification 静态机械放大倍数
static correction 静校正
static positioning 静态定位
static sensor 静态遥感器
station 测站

stationary front 静止锋
stationary meteor 驻留流星
stationary orbit 定常轨道
stationary prominence flare 稳定日珥型耀斑
stationary radiant 固定辐射点
station centring 测站归心
statistical error 统计误差
statistical fluctuation 统计涨落
statistical mechanics 统计力学
statistic map 统计地图
steady-state creep 稳态蠕变
steady-state theory 稳态理论
steaming ground 冒汽地面
steam vent 汽孔
steep incidence 陡峭入射
Stefan's law 斯蒂芬定律
stellar camera 恒星摄影机
stellar parallax 恒星视差
stepout 时差
stereocamera 立体摄影机
stereocomparator 立体坐标量测仪
stereographic projection 球面投影
stereointerpretoscope 立体判读仪
stereometric camera 立体摄影机
stereopair 立体像对
stereophotogrammetry 立体摄影测量
stereoplotter 立体测图仪
stereoscope 立体镜
stereoscopic map 视觉立体地图
stereoscopic model 立体观测模型
stereoscopic observation 立体观测
stereoscopic vision 立体视觉
stick plot 短棒图
stick slip 黏滑
stick-up lettering 透明注记
stiffness 劲度
stipple 网点
stochastic acceleration 随机加速
Stoneley wave 斯通莱波
stony desert 石漠
stope survey 采场测量

stopping cross section 阻止截面
stopping phase 停止相
storm after-effect 暴后效应
Stormer cone 斯特默锥
Stormer length 斯特默长度
storm morphology 磁暴形态学
storm sudden commencement 磁暴急始
storm-time proton belt 暴时质子带
storm-time variation 暴时变化
strain accumulation 应变积累
strainmeter 应变仪
strain step 应变阶跃
stratopause 平流层顶
stratosphere 平流层
stratosphere coupling 平流层耦合
stratospheric circulation 平流层环流
stratospheric dynamics 平流层动力学
stratospheric inversion layer 平流层逆温层
stratospheric structure 平流层结构
stratospheric thermal field 平流层热场
stratospheric variation 平流层变化
stratospheric wind field 平流层风场
streamer 海洋地震拖缆
streaming aurora 流动状极光
streaming plasma 流动等离子体
streamline analysis 流线分析
stream of fast plasma 快速等离子体流
strength of the ring current 环电流强度
stress dislocation 应力位错
stress drop 应力降
stress glut 应力过量
stressmeter 应力仪
stress relief 应力解除
stress tensor 应力张量

stress trajectory 应力迹线
stretched-out field 伸展磁场
strike orientation 走向定向
stripping and mining engineering profile 采剥工程断面图
stripping the Earth 剥地球(法)
strong (ground) motion 强地面运动,强地动
strong coupling 强耦合
strong earthquake 强震
strong frontal zone 强锋带
strong geomagnetic activity 强地磁活动
Strong-Motion Array in Taiwan Number 1 (SMART 1) 台湾强地动一号台阵
strong-motion seismograph 强震仪
strong-motion seismology 强地动地震学
strontium cloud 锶云
structural geochemical anomaly 构造地球化学异常
stud registration 销钉定位法
subauroral latitude 副极光带纬度
subauroral region 副极光区
subauroral zone 副极光带,亚极光带
sub-bottom profiler 浅地层剖面仪
subcrustal earthquake 壳下地震
subdivisional organization 再分结构
subduction 消减
subduction belt 消减带
subduction-type geothermal belt 消减型地热带
subduction zone 消减带
subequatorial belt 副赤道带
sublunar point 月下点
submarine earthquake 海下地震
submarine fumarole 洋底喷气孔
submarine hot spring 洋底热泉
submarine seismograph 海底地震仪

subordinate geochemical landscape 从属的地球化学景观
subpolar zone ionosphere 副极地带电离层
sub-satellite point 卫星星下点
sub-solar density bulge 日下点密度隆起
sub-solar point 日下点
substorm activity 亚暴活动
substorm current 亚暴电流
substorm intensity 亚暴强度
substratosphere 副平流层
subsurface grid 地下网格
sub-visual aurora 亚视阈极光
successive approximation 逐步接近
successive contrast 连续对比
successive injection 连续注入
successive sheet model 逐层模型
sudden commencement 急始
sudden commencement (magnetic) storm 急始磁暴
sudden frequency deviation (SFD) (短波)频率急偏
sudden impulse 急脉冲
sudden increase of ionization 电离突增
sudden ionospheric disturbance (SID) 突发电离层骚扰
sudden onset 急始
sudden phase anomaly (SPA) 突发相位异常
sudden short wave fade-out 短波突然衰退
sudden stratospheric warming 平流层突然增温
suitability 适宜性
summer drought 夏旱
summer solstice 夏至(点)
sun compass 太阳罗盘
sun-earth line 日地线
sunlit aurora 日照极光
sunlit barium cloud 日照钡云
sun proton monitor 太阳质子监视仪
sunspot activity 黑子活动性

sunspot cycle 黑子周
sunspot cycle variation 黑子周变化
sunspot flare 黑子耀斑
sunspot maximum 黑子极大值
sunspot minimum 黑子极小值
sunspot polarity 黑子极性
sunspot prominence 黑子日珥
sunspot radiation 黑子辐射
sun's total radiation 太阳总辐射
sun-synchronous 太阳同步
sunward side of the earth 地球向阳侧
sunward tail 向阳彗尾
superconductive gravimeter 超导重力仪
superconductive magnetometer (SQUID magnetometer) 超导磁力仪
superconductivity 超导性
super-corona 超日冕
superior planet 外行星
superposed-epoch analysis 时间叠加分析
superrotational 超旋转的
supersonic expansion 超声膨胀
supersonic polar wind 超声极风
supersonic solar wind 超声速太阳风
super stratosphere 超平流层
super-thermal photoelectron 过热光电子
supervised classification 监督分类
suppression 压制
surface-borehole variant 地面-井中方式
surface brightness 表面亮度
surface charge 表面电荷
surface geochemical survey 地表地球化学测量
surface geothermal manifestation 地表地热显示
surface gravity 表面重力
surface heat flow 地表热流
surface hodograph 时距曲面

surface pressure 地面气压
surface soil 表层土壤
surface S wave 横波型面波
surface-underground contrast plan 井上下对照图
surface wave 面波
surface wave magnitude 面波震级
survey adjustment 测量平差
survey for land smoothing 平整土地测量
survey for marking of boundary 标界测量
survey grid 测网
surveying 测量学
surveying and mapping (SM) 测绘学
surveying and mapping engineering 测绘工程
surveying control network 测量控制网
surveying for site selection 厂址测量
survey in mining panel 采区测量
survey in reconnaissance and design stage 勘测设计阶段测量
survey line 测线
survey mark 测量标志
survey of existing station yard 既有线站场测量
survey station 测点, 测站
survey vessel 测量船
susceptibility 磁化率
swamp seismic exploration 沼泽地震勘探
swarm 震群
swath 测线束法
s-wave S波
sweep frequency sounder 扫频测深仪
SWF (short wave fade-out) 短波衰退
swing angle 像片旋角
Sycorax 天卫十七
symbol 符号
symbolization 符号化

symmetrical four-pole sounding 对称四极测深
symmetrical mode 对称振型
symmetrical profiling 对称剖面法
sympathetic radio burst 共振射电爆发
synchronous altitude 同步高度
synchronous changes 同步变化
synchronous orbit 同步轨道
synchronous rotation 同步旋转
synchrotron radiation 同步辐射
syngenetic geochemical anomaly 同生地球化学异常
synodic lunar period 会合太阴周
synodic motion 会合运动
synodic rotation 会合自转
synodic satellite 会合卫星
synoptic chart 天气图
synoptic observation 天气观测
synthetical seismogram 合成地震图
Synthetic Aperture Radar (SAR) 合成孔径雷达
synthetic map 合成地图
synthetic plan of stripping and mining 采剥工程综合平面图
systematic error 系统误差

tactual map 触觉地图
tadpole plot 蝌蚪图
tail stingers system 尾刺系统
take-off angle 离源角
tangential distortion 切向畸变
tangential lens distortion 切向畸变
tangent off-set method 切线支距法
target 觇牌
target area 目标区
target reflector 目标反射器
target road engineering survey 靶道工程测量
T-axis (tension axis) 张力轴，T轴
TEC (total electron content) 电子总含量
tectonic activity 构造活动
tectonic earthquake 构造地震
tectonic stress 构造应力
tectonophysics 构造物理学
telemetered seismic network 遥测地震台网
telemetric seismic instrument 遥测地震仪
telemetry 遥测
telemetry system 遥测系统
teleseism 远震
teleseismic wave 远震地震波
Telesto 土卫十三
telluric (current) method 大地电流法
telluroid 近似地面形
temperate latitude 中纬度
temperature change 温度变化
temperature effect 温度效应

temperature logging 温度测井
temperature profile 温度剖面
temporal regularity 时序规则性
tensile dislocation 张位错
tension axis (T-axis) 张力轴，T轴
tension in the tail 磁尾张力
tenuous plasma 稀薄等离子体
terrain 地形
terrain correction 地形校正
terrestrial effect 地球效应
terrestrial event 地球事件
terrestrial heat flow 大地热流
terrestrial interferometry 地球干涉量度学
terrestrial meridian 地球子午线
terrestrial parallel 地面纬圈，地球纬度圈
terrestrial radiation 地球辐射
terrestrial radiation flux 地球辐射通量
terrestrial space 近地空间
terrestrial spectroscopy 地球谱学
territorial sea baseline survey 领海基线测量
tertiary creep 第三期蠕变
Tethys 土卫三
thematic atlas 专题地图集
thematic cartography 专题地图学
thematic map 专题地图
thematic mapper 专题制图仪
thematic overlap 专题层
theodolite 经纬仪
theodolite traverse 经纬仪导线
theoretical astronomy 理论天文学

theoretical astrophysics 理论天体物理学
theoretical cartography 理论地图学
theoretical geochemistry 理论地球化学
theoretical seismogram 理论地震图
theory of relativity 相对论
theory of similarity 相似理论
thermal anisotropy 热各向异性
thermal cleaning 热清洗
thermal diffusion 热扩散
thermal effect 热效应
thermal energy density 热能密度
thermal energy distribution 热能分布
thermal equilibrium 热平衡
thermal escape 热逃逸
thermal expansion 热膨胀
thermal ionization 热致电离
thermally-driven tide 热致潮汐
thermal plasma 热等离子体
thermal plasma drift 热等离子体漂移
thermal radiation 热辐射
thermal tide motion 热致潮汐运动
thermodynamical function 热力学函数
thermodynamic process 热力学过程
thermomagnetic curve 热磁曲线
thermomagnetic separation 热磁分离
thermopause 热层顶
thermoremanence 热剩磁
thermoremanent magnetization (TRM) 热剩磁
thermosphere 热层
thermospheric event 热层事件
thin emitting layer 薄发射层
third cosmic velocity 第三宇宙速度
third invariant 第三不变量
Thomson-Haskell matrix methord 汤姆森-哈斯克尔矩阵法
Thomson scattering 汤姆森散射
three-body attachment 三体附着
three-body reaction 三体反应
three-body recombination 三体复合
three-dimensional data volume 三维数据体
three-dimensional display 三维显示
three-dimensional migration 三维偏移
three-dimensional seismic method 三维地震法
three-dimensional terrain 三维地形
three-dimensional terrain simulation 三维地景仿真
three-dimensional visualization 三维可视化
threefold coincidence 三重符合
threshold energy 阈能
threshold rigidity 阈值刚度
thunderstorm electricity 雷暴电学
thunderstorm event 雷暴事件
thunderstorm flash 雷暴闪光
Thvera event 斯韦劳事件
TID (travelling ionospheric disturbance) 电离层行扰
tidal 潮汐的
tidal action 潮汐作用
tidal effect 潮汐效应
tidal factor 潮汐因子
tidal harmonic analysis 潮汐调和分析
tidal harmonic constant 潮汐调和常数
tidal information panel 潮信表
tidal motion 潮汐运动
tidal nonharmonic analysis 潮汐非调和分析
tidal nonharmonic constant 潮汐非调和常数
tidal observation 验潮
tidal oscillation 潮汐振荡
tidal perturbation 潮汐摄动

tidal prediction 潮汐预报
tidal station 验潮站
tidal table 潮汐表
tidal wave 潮汐波
tidal zone seismic 潮汐带地震勘探
tide-generating force 引潮力
tide-generating potential 引潮位
tide-meter 验潮仪
tie point 连结点
tillage 耕作
tilt 倾斜
tilt angle of photograph 像片倾角
tilt correction 倾斜改正
tilt displacement 倾斜位移
tiltmeter 倾斜仪
tilt observation 倾斜观测
time (record) section 时间剖面
time break 爆炸信号
time depth conversion 时深转换
time derivative 时间导数
time-distance curve (T-X curve) 时距曲线
time of reversal 倒转时间
time pulse 时间脉冲
time resolution 时间分辨率
time signal in UTC 协调世界时时号
time slice 时间切片
time smoothing 时间平滑
time-term 时间项
time-term method 时间项法
time-variable filtering 时变滤波
time variant scaling 时变比例
timing marker 时标,时号
Tiros satellite 泰罗斯卫星
Titan 土卫六
Titania 天卫三
TM image TM影像
tolerance 限差
tomographic 层析
tomography 层析成像
tone 色调
topocentric coordinate system 站心坐标系

topographic control point 地形控制点
topographic correction 地形校正
topographic database 地形数据库
topographic diagram 位形图
topographic feature 地貌
topographic feature survey 地貌调查
topographic isobar 地形等压线
topographic isotopic fractionation effect 地形同位素分馏效应
topographic map 地形图
topographic map of bridge site 桥位地形图
topographic map of mining area 井田区域地形图
topographic map of urban area 城市地形图
topographic Rossby wave 地形罗斯贝波
topography 地形
topological 拓扑的
topological map 拓扑地图
topological relation 拓扑关系
topological retrieval 拓扑检索
topside ionogram 俯测频高图
topside ionosphere 顶外电离层
topside sounder 顶视探测仪
tornado prominence 龙卷日珥
toroidal 环型
toroidal field 环形场
toroidal oscillation 环型振荡
torsional 扭转型
torsional oscillation 扭转型振荡
total absorption 总吸收
total accuracy of sounding 测深精度
total-annular eclipse 全环食
total drag acceleration 总阻力加速度
total eclipse 全食
total electron content (TEC) 电子总含量
total intensity of magnetic anomaly 总磁异常强度

total ion flux 总离子通量
total lunar eclipse 月全食
total nitrogen 全氮
total ozone amount 总臭氧量
total radiation 全辐射
total reflection 全反射
total soft electron flux 软电子总通量
total solar eclipse 日全食
total thermoremanent magnetization 总热剩磁
tourist map 旅游地图
toward polarity 向阳极性
toward sector 向阳扇区
towed bird system 吊舱系统
towed boom 托架
T phase T震相
trace constituent 痕量成分,微量成分
trace equalization 道间均衡
tracing 追踪
tracing digitizing 跟踪数字化
track detecter 径迹探测器
tracking 跟踪
track of electron 电子径迹
track station 基准台,差分台
trade-off studies 折中选择研究
trajectory measuring system 测轨系统
trajectory of fluid motion 流体运动轨迹
transducer 换能器
transducer baseline 换能器基线
transducer dynamic draft 换能器动态吃水
transducer static draft 换能器静态吃水
transfer ellipse 转移椭圆
transfer function 传递函数
transfer orbit 转移轨道
transformation 转换
transform fault 转换断层
transient 瞬变
transient change 瞬时变化
transient creep 暂态蠕变
transient event 瞬时事件

transient field method 瞬变场法,过渡场法
transient plasma event 瞬时等离子体事件
transitional zone seismic 过渡带地震勘探
transition curve location 缓和曲线测设
transition layer 过渡层
transition probability 跃迁概率
transit path 渡越路径
translation parameter 平移参数
transmission coefficient 透射系数
transmission function 透射函数
transmission matrix 透射矩阵
transmitting line of sounder 测深仪发射参数,测深仪零线
transparent foil 网纹片
transparent plasma 透明等离子体
transport 输移
transport equation 输运方程
transverse conductivity 横向电导率
transverse drift 横向漂移
transverse projection 横轴投影
transverse propagation 横向传播
transverse wave 横波
trap 圈闭
trapped energy 捕获能量
trapped particle 捕获粒子
trapped particle dynamics 捕获粒子动力学
trapping phenomenon 捕获现象
travelling disturbance 运动性扰动
travelling ionospheric disturbance (TID) 电离层行扰
travelling wave 行波
travel time （地震波）走时
travel-time curve 走时曲线
travel-time table 走时表
traverse 导线
tree-ring climatology 年轮气候学
triangulation 三角测量
triangulation chain 三角锁

triangulation network 三角网
triangulation point 三角点
triaxial magnetometer 三轴磁强计
trigger action 触发作用
trigger effect 触发效应
triggering 触发
trigger mechanism 触发机制
trigonometric 三角学的
trigonometric leveling 三角高程测量
trigonometric leveling network 三角高程网
Triton 海卫一
TRM (thermoremanent magnetization) 热剩磁
tropical 热带的
tropical airglow 热带气辉
tropical cyclone 热带气旋
tropical red arc 热带红弧
tropical troposphere 热带对流层
tropopause 对流层顶
troposphere 对流层
tropospheric event 对流层事件
trough 海槽
trough-like structure 槽式结构
true error 真误差
true height 真高
true horizon 真地平线,真水平线
true meridian 真子午线
true reflection height 真反射高度
truncation error 截断误差
tsunami 海啸
tsunami earthquake 海啸地震
tunneling effect （地震波的）隧道效应
tunneling effect of seismic waves 地震波的隧道效应
tunneling wave 隧道波
Turam method 土拉姆法

turbidity factor 浑浊因子
turbopause 湍流层顶
turbosphere 湍流层
turbulence 湍流,紊流
turbulence component 湍流分量
turbulence spectrum 湍流谱
turbulent boundary layer 湍流边界层
turbulent diffusion 湍流扩散
turbulent dissipation 湍流耗散
turbulent exchange 湍流交换
turbulent magnetic field 湍流磁场
turbulent mixing 湍流混合
turbulent motion 湍流运动
turbulent plasma 湍流等离子体
turbulent transfer 湍流输送
turning point 转折点
twilight 曙暮光,晨昏蒙影
twilight airglow 曙暮气辉
twilight arc 曙暮光弧
twilight colour 曙暮霞
twilight emission 曙暮光发射
twilight spectrum 曙暮光谱
twinkling map 瞬间地图
twisted flux tube 扭转通量管
two-component plasma 二元等离子体
two-dimensional dipole model 二维偶极模型
twofold coincidence 双重符合
two-stream instability 双流不稳定性
two-stream plasma 双流等离子体
T-X curve (time-distance curve) 时距曲线
typal map 类型地图
typical reaction time 典型反应时间

UGIS (urban geographical information system) 城市基础地理信息系统
ultra-high frequency 超高频
ultrametamorphic water 超变质水
ultrasonic image logging 超声成像测井
ultraviolet astronomy 紫外天文学
ultraviolet dayglow 紫外日气辉
ultraviolet flux 紫外通量
ultraviolet ion source 紫外离子源
ultraviolet radiation 紫外辐射
ultraviolet spectrum 紫外光谱
umbra 本影
umbral eclipse 本影食
umbral flash 本影闪烁
Umbriel 天卫二
unblocking field 解阻场
unblocking temperature 解阻温度
uncertainty principle 不确定原理
under colour addition 底色增益
under colour removal 底色去除
underground cavity survey 井下空硐测量
underground fluid 地下流体
underground survey 井下测量,矿井测量
underthrust belt 俯冲带
undisturbed solar condition 未扰动太阳状况
unidirectional flux 单向通量
unified magnitude 统一震级
uniform electric field 均匀电场
uniformly magnetized sphere 均匀磁化球
unilateral faulting 单侧断裂
unimodal distribution 单峰分布
unipolar group 单极群
unipolar magnetic region 单极磁区
unipolar sunspot 单极黑子
universal abundance 宇宙丰度
universal method of photogrammetric mapping 全能法测图
Universal Polar Stereographic projection (UPS) 通用极球面投影
Universal Transverse Mercator projection (UTM) 通用横墨卡托投影
unmagnetized plasma 非磁化等离子体,无磁等离子体
unperturbed atmosphere 未扰动大气
unperturbed model 未扰动模式
unsteady 非恒定的
upgoing electron 上升电子
uphole time 井口时间
upper air observation 高空观测
upper atmosphere 高层大气
upper atmospheric physics 高层大气物理学
upper chromosphere 色球高层
upper envelope 上包络
upper hybrid frequency 高混杂频率
upper level streamline 高空流线
upper mantle 上地幔
upper mesosphere 高中层
upper tropical troposphere 热带高对流层

UPS (Universal Polar Stereographic projection) 通用极球面投影
upstreaming ion 上流离子
upstream shock 上流激波
up sweep 升频扫描
up-to-date map 现势地图
upward ion current 上流离子流
Uranus 天王星
urban control survey 城市控制测量
urban geographical information system (UGIS) 城市基础地理信息系统
urban land 城市土地
urban landscape 城市景观
urban mapping 城市制图
urban survey 城市测量
urban topographic survey 城市地形测量
UTC (coordinate universal time) 协调世界时
UTM (Universal Transverse Mercator projection) 通用横墨卡托投影

Van Allen belt 范艾仑(辐射)带
vanishing line 消失线；合线；没影线
vanishing point control 合点控制
variable-frequency method 变频法
variomat 变线仪
variometer 磁变仪
varioscale projection 变比例投影
vectograph method of stereoscopic viewing 偏振光立体观察
vector 矢量
vector data 矢量数据
vector image 矢量图形
vectorization 矢量化
vector plotting 矢量绘图
vegetation 植被
vegetation restoration 植被恢复
vegetation type 植被类型
velocity 速度
velocity dispersion 速度弥散度
velocity field 速度场
velocity filtering 速度滤波
velocity gradient 速度梯度
velocity of escape 逃逸速度
velocity potential 速度势
velocity response spectrum 速度反应谱
velocity structure 速度结构
Vening Meinesz isostasy 韦宁迈内兹均衡
Venus 金星
Venus circulation 金星环流
Venus seismology 金星震学
vertical 纵向的
vertical circle 地平经圈，垂直圈
vertical coaxial coils system 垂直同轴线圈系统
vertical component 垂直分量
vertical coplanar coils system 垂直共面线圈系统
vertical density profile 垂直密度剖面
vertical distribution 垂直分布
vertical drift 垂直漂移
vertical energy flux 垂直能量通量
vertical epipolar line 垂核线
vertical epipolar plane 垂核面
vertical flux 垂直通量
vertical incidence 垂直入射
vertical instability 垂直不稳定性
vertical intensity 垂直强度
vertical optical thickness 垂直光学厚度
vertical ozone profile 臭氧垂直剖面
vertical pressure gradient 垂直气压梯度
vertical refraction coefficient 垂直折光系数
vertical refraction error 垂直折光误差
vertical section 垂直截面
vertical seismic profile 垂直地震剖面
vertical seismic profile log 垂直地震剖面测井
vertical seismic profiles survey (VSP survey) 垂直地震测线法
vertical shear 垂直切变
vertical shear vector 垂直切变矢量
vertical stacking 垂直叠加

vertical transport 垂直输送
very high frequency 甚高频
very long baseline interferometry (VLBI) 甚长基线干涉测量
very low frequency 甚低频
very low frequency band radiated field system 甚低频带辐射场系统
very low frequency method (VLF method) 甚低频法
VGP (virtual geomagnetic pole) 虚地磁极
vibrational level 振动能级
vibrational transition 振动跃迁
vibroseis 可控震源法
video frequency 视频
virgin stress 初始应力
virtual displacement 虚位移
virtual geomagnetic pole (VGP) 虚地磁极
virtual height 虚高
virtual landscape 虚拟地景
virtual map 虚拟地图
virtual reflection height 虚反射高度
viscous coupling 黏度耦合
viscous force 黏滞力
viscous-like interaction 黏滞性相互作用
viscous remanence 黏滞剩磁
viscous remanent magnetization (VRM) 黏滞剩磁
visibility 能见度
visibility acuity 能见敏锐度
visible aurora 可见极光
visible radiation 可见辐射
visible spectrum 可见光谱
visual aurora 目视极光
visual balance 视觉平衡
visual contrast 视觉对比
visual hierarchy 视觉层次
visual interpretation 目视判读
visualization 可视化

visualization of spatial information 空间信息可视化
visualization research 可视化研究
visualization technology 可视化技术
visual meteor 目视流星
visual observer 目视观测者
visual variable 视觉变量
visual zenith telescope 目视天顶仪
Vlaslov equation 伏拉索夫方程
VLBI (very long baseline interferometry) 甚长基线干涉测量
VLF emission 甚低频发射
VLF method (very low frequency method) 甚低频法
VLF noise 甚低频噪声
volcanic activity 火山活动
volcanic earthquake 火山地震
volcanic mound seismic facies 火山丘地震相
volcanic rock 火山岩
volcanic water 火山水
volcanism 火山作用
volcano 火山
volcano-geothermal region 火山地热区
volume charge 体积电荷
volume element 体积元
volume emission 体发射
vorticity 涡度,涡量
vorticity equation 涡度方程
vorticity field 涡度场
vorticity transfer 涡度传输
vorticity transport theory 涡度输送理论
VRM (viscous remanent magnetization) 黏滞剩磁
VSP survey (vertical seismic profiles survey) 垂直地震测线法

WACDP (wide angle common depth point) 广角共深度点
Wadati diagram 和达图
walkaway seismic profiling 逐点激发地震剖面法
walled plain 月面环壁平原，环形低地
wall effect 壁效应
warm front 暖锋
warm plasma 暖等离子体
water absorption spectrum 水吸收光谱
water and soil 水土
water bottom event 海底波
water head 水头
water ionization limit 水电离限
water level 水位
water radon 水氡
water saturation 含水饱和度
watershed hydrology 流域水文学
water surface evaporation 水面蒸发
water system 水系
wave beam angle 波束角
wave crest 波峰
wave disturbance 波状扰动
wave energy transport 波能输送
wave equation 波动方程
wave equation migration 波动方程偏移
wave field 波场
waveform 波形
wave frequency 波动频率
wave front 波阵面，波前
wave guide 波导
wave impedance 波阻抗
wavelength 波长
wavelet 小波，子波
wavelet analysis 小波分析
wavelet processing 子波处理
wavelet transform 小波变换
wave mechanics 波动力学
wave noise 波噪声
wave normal 波面法线
wave packet 波包
wave-particle interaction 波粒相互作用
wave propagation 波传播
wave spectrum 波谱
wave speed 波速
wave trough 波谷
wave vector 波矢量
wave velocity 波速
wave-wave interaction 波波相互作用
wedge seismic facies 楔状地震相
weighted stack 加权叠加
weight reciprocal of figure 图形权倒数
well fluid logging 井液测井
well logging 测井
well shooting 地震测井
well zone 井区
Wenner array 温纳排列
Wentzel-Kramers-Brillouin-Jeffreys method (WKBJ method) WKBJ法
wetland 湿地
wet model 湿模式
whistler 哨声
whistler dispersion 哨声波散
white light emission 白光发射
white light event 白光事件
white noise 白噪音

wide angle common depth point (WACDP) 广角共深度点
wide-angle object glass 广角物镜
wide-angle reflection 大角度反射
wide-angle telescope 广角望远镜
wide line profile 宽线剖面
wide line seismic 宽线地震
Wiechert seismograph 维歇特地震仪
Wien's law 维恩定律
Willmore seismograph 威尔莫地震仪
wind erosion 风蚀
windowing 开窗
wind shear 风切变
wind vector 风矢量
windward 迎风的
windward side 迎风面
wing-tip system 翼梢系统
winter anomaly 冬季异常
winter hemisphere 冬半球
winter solstice 冬至
within-sites precision 采点内精度

WKBJ seismogram WKBJ 地震图
WKBJ theoretical seismogram WKBJ 理论地震图
WKBJ method (Wentzel-Kramers-Brillouin-Jeffreys method) WKBJ 法
Wolf number 沃尔夫(黑子)数
Wood-Anderson seismograph 伍德-安德森地震仪
woodland 林地
woodland soil 林地土壤
workflow 工作流
work function 功函数
world atlas 世界地图集
world climate 世界气候
worldwide aurora 全球极光
worldwide potential gradient 全球电位梯度
World Wide Standard Seismograph Network (WWSSN) 世界范围标准地震台网
WWSSN (World Wide Standard Seismograph Network) 世界范围标准地震台网

xerography 静电复印
X-ray X 射线
X-ray astronomy X 射线天文学
X-ray background X 射线背景
X-ray burst X 射线暴
X-ray detector X 射线探测器

X-ray flux X 射线通量
X-ray optics X 射线光学
X-ray radiation X 射线辐射
X-ray substorm X 射线亚暴
X-ray telescope X 射线望远镜
X-source X 射线源

yacht chart 游艇用图
yaw 偏航
yaw angle 偏航角

yawing axis 偏航轴
yield function 产额函数

Zeeman effect 塞曼效应
zenith 天顶
zenithal arc 天顶光弧
zenithal projection 天顶投影
zenith angle 天顶角
zenith distance 天顶距
zenith telescope 天顶仪
zenographic 木星表面的
zero flow velocity 零流速
zero-frequency seismology 零频地震学
zero-g (zero gravity) 失重,零重力
zero gravity (zero-g) 失重,零重力
zero-initial-length spring 零长弹簧
zero meridian 零子午线
zero-phase effect 零相位效应
zero point of the tidal 验潮站零点
zero zone 零时区
Zijderveld diagram 泽德费尔德图
Z-number 原子序数
zodiac 黄道带
zodiacal circle 黄道圈
zodiacal light 黄道光
Zollner suspension 措尔纳悬挂法
zonal circulation 纬向环流
zonal kinetic energy 纬向动能
zonal wave number 纬向波数
zonal wind 纬向风
zone of eclipse 食带
zone of totality 全食带

非英文字母开头的词条

20°discontinuity 20°间断
3-D data volume 三维数据体
3-D display 三维显示
3-D migration 三维偏移
3-D seismic method 三维地震法
3-D terrain 三维地形
3-D terrain simulation 三维地景仿真
3-D visualization 三维可视化
α-ray α射线
α-track etch survey α径迹测量
β-ray β射线
γ-neutron method γ-中子法
γ-ray logging 自然γ测井
γ-ray radiation γ射线辐射
γ spectrometer γ能谱仪
γ-γ logging γγ测井
τ function τ函数
τ method τ法

汉英部分

阿贝比长原理 Abbe comparator principle
阿达马变换 Hadamard transformation
阿尔文层 Alfvón layer
阿尔文马赫数 Alfvón Mach number
阿尔文微扰理论 Alfvón perturbation theory
阿佛伽德罗数 Avogadro number
阿基米得螺线 Archimedes spiral
阿普尔顿层 Appleton layer
阿普尔顿-哈特里公式 Appleton-Hartree formula
阿普尔顿异常 Appleton anomaly
埃达台网 IDA Network (International Deployment of Accelerometers Network)
艾里-海斯卡宁均衡 Airy-Heiskanen isostasy
艾里震相 Airy phase
爱丁顿极限 Eddington limit
安平精度 setting accuracy
氨 ammonia
氨电离限 ammonia ionization limit
氨基酸地球化学法 amino acid geochemistry method
岸台 base station
暗点 dim spot
暗礁 reef
暗谱斑 dark flocculus
凹凸体 asperity
凹凸体震源模式 asperity source model
奥杜瓦伊事件 Olduvai event
奥格簇射 Auger shower
奥陶系 ordovician system
奥陶系的 ordovician

Bb

八极 octupole
巴耳末公式 Balmer formula
巴耳末间断 Balmer discontinuity
巴耳末跳跃 Balmer discontinuity
靶 barn
靶道工程测量 target road engineering survey
白光发射 white light emission
白光事件 white light event
白噪音 white noise
百分频率效应 percent frequency effect
斑块 patch
板间地热带 interplate geothermal belt
板间地震 interplate earthquake
板块 plate
板块大地构造学 plate tectonics
板块构造学 plate tectonics
板块碰撞 plate collision
板内地热系统 intraplate geothermal system
板内地震 intraplate earthquake
板内火山 intraplate volcano
半导体激光器 semiconductor laser
半功率点 half-power point
半距等高线 half-interval contour
半年效应 semi-annual effect
半日变化 semidiurnal variation
半日潮港 semidiurnal tidal harbour
半日潮汐分量 semidiurnal tidal component
半色调 halftone
半影 half-shadow
半影月食 appulse
半影锥 penumbra cone
半子午线 semi-meridian
邦德反照率 Bond albedo
饱和度 saturation
暴后效应 storm after-effect
暴时变化 storm-time variation
暴时质子带 storm-time proton belt
曝光 exposure
爆发簇射 explosive shower
爆发拱 eruptive arch
爆发日珥 eruptive prominence
爆发源 burst source
爆炸地震学 explosion seismology
爆炸信号 time break
爆炸震源 explosive source
北极 Arctic
北极盖 northern polar cap
北极光 aurora borealis; aurora polaris; northern lights
北极圈 Arctic Circle
北极星任意时角法 method by hour angle of Polaris
贝尼奥夫带 Benioff zone
贝尼奥夫地震仪 Benioff seismograph
贝塞尔大地主题解算公式 Bessel formula for solution of geodetic problem
贝塞尔方程 Bessel equation
贝塞尔函数 Bessel function
贝塞尔椭球 Bessel ellipsoid
贝叶斯分类 Bayesian classification
背阳半球 night hemisphere

背阳极性　away polarity
背阳扇区　away sector
钡云　barium cloud
钡蒸汽　barium vapour
倍增过程　multiplicity process
倍增因子　multiplicity factor
被动式遥感　passive remote sensing
被动源法　passive source method
被动源方法　passive source method
本初子午线　prime meridian
本身亮度　intrinsic brightness
本影　umbra
本影闪烁　umbral flash
本影食　umbral eclipse
本征值　eigen value
崩溃　breakdown
泵站　pumping station
鼻哨　nose whistler
鼻形结构　nose structure
比尔-兰伯特定律　Beer-Lambert law
比较地图学　comparative cartography
比例尺　scale
比例量表　ratio scaling
比例误差　proportional error
比相　phase comparison
比值变换　ratio transformation
比值增强　ratio enhancement
比重　specific gravity
彼得森电导率　Pedersen conductivity
币形裂纹　penny-shaped crack
闭合　closure
闭合差　closing error; closure
闭合导线　closed traverse
闭合回线场　closed loop field
闭合漂移轨道　closed drift orbit
闭合水准路线　closed leveling line
闭锁断层　locked fault
壁效应　wall effect
边长中误差　mean square error of side length

边交会法　linear intersection
边角交会法　linear-angular intersection
边界条件　boundary condition
边坡　side slope
边缘检测　edge detection
边缘增强　edge enhancement
编绘　compilation
编绘原图　compiled original
编制　compilation
变比例投影　varioscale projection
变幅指数　range index
变换光束测图　affine plotting
变频法　variable-frequency method
变线仪　variomat
变形观测控制网　control network for deformation observation
变形椭圆　indicatrix ellipse
变质水　metamorphic water
标称精度　nominal accuracy
标乘　scalar multiplication
标尺　rod; staff
标定　calibration
标定球　calibration sphere
标高　scale height
标高差改正　correction for skew normals
标界测量　survey for marking of boundary
标量　scalar quantity
标量乘法　scalar multiplication
标量磁强计　scalar magnetometer
标量强度　scalar intensity
标志层　key bed
标志灯　signal lamp
标准层　key bed
标准高度分析　standard-height analysis
标准配置点　Gruber point
标准偏差　standard deviation
标准频率　standard frequency
标准纬线　standard parallel

标准误差 standard error
标准子午线 standard meridian
表层土壤 surface soil
表面电荷 surface charge
表面亮度 surface brightness
表面重力 surface gravity
冰川 glacier
冰川时代 glacial epoch
冰川学 glaciology
冰后回弹 post glacial rebound
冰后期 post-glacial time
冰期 ice age
波包 wave packet
波波相互作用 wave-wave interaction
波长 wavelength
波场 wave field
波传播 wave propagation
波茨坦重力系统 Potsdam gravimetric system
波导 wave guide
波动方程 wave equation
波动方程偏移 wave equation migration
波动力学 wave mechanics
波动频率 wave frequency
波峰 wave crest
波谷 wave trough
波浪补偿 compensation of undulation; heave compensation
波浪补偿器 heave compensator
波粒相互作用 wave-particle interaction
波罗-科普原理 Porro-Koppe principle
波面法线 wave normal
波能输送 wave energy transport
波谱 wave spectrum
波谱测定仪 spectrometer
波谱集群 spectrum cluster
波谱特征空间 spectrum feature space
波谱特征曲线 spectrum character curve
波谱响应曲线 spectrum response curve
波前 wave front
波矢量 wave vector
波束角 beam angle; wave beam angle
波束天线 beam antenna
波束效率 beam efficiency
波速 wave speed; wave velocity
波形 waveform
波噪声 wave noise
波阵面 wave front
波状扰动 wave disturbance
波阻抗 wave impedance
玻耳兹曼常数 Boltzmann constant
玻耳兹曼公式 Boltzmann formula
玻什-大森地震仪 Bosch-Omori seismograph
玻意耳定律 Boyle's law
剥地球(法) stripping the Earth
伯格反褶积 Burg deconvolution
伯克兰电流 Birkeland current
伯努利定理 Bernoulli's theorem
泊松方程 Poisson's equation
泊松分布 Poisson distribution
泊位 berth
博姆判据 Bohm criterion
薄发射层 thin emitting layer
补偿器 compensator
补偿器补偿误差 compensating error of compensators
补偿深度 depth of compensation
补偿线性向量偶极 CLVD (compensated linear vector dipole)
捕获 acquisition
捕获和跟踪雷达 acquisition and tracking radar
捕获粒子 trapped particle
捕获粒子动力学 trapped particle dynamics
捕获能量 trapped energy
捕获现象 trapping phenomenon
捕捉截面 capture cross-section
不闭合漂移轨道 open drift orbit

不变地磁坐标 invariant geomagnetic coordinate
不变纬度 invariant latitude
不规则地磁场 irregular geomagnetic-field
不规则极光斑 irregular auroral patch
不均匀的 inhomogeneous
不均匀介质 inhomogeneous medium
不连续束 discrete beam
不确定原理 uncertainty principle
不透明等离子体 opaque plasma
布尔诺漂移 Brno excursion
布格改正 Bouguer correction
布格校正 Bouguer reduction
布格异常 Bouguer anomaly
布莱克漂移 Blake excursion
布莱克事件 Blake event
布朗热改正 Browne correction
布利登指数 Briden index
布隆斯公式 Bruns formula
布容期 Brunhes epoch
布耶哈马问题 Bjerhammar problem
部分热剩磁 PTRM (partial thermoremanent magnetization)
部分无滞剩磁 PARM (partial ARM)

采剥工程断面图 stripping and mining engineering profile
采剥工程综合平面图 synthetic plan of stripping and mining
采场测量 stope survey
采间点精度 between-sites precision
采点内精度 within-sites precision
采掘工程平面图 mining engineering plan
采区测量 survey in mining panel
采区联系测量 connection survey in mining panel
采样 sampling
采样间隔 sampling interval
彩色编码 colour coding
彩色变换 colour transformation
彩色复制 colour reproduction
彩色感光材料 colour sensitive material
彩色红外片 colour infrared film
彩色校样 colour proof
彩色片 colour film
彩色摄影 colour photography
彩色线划校样 dye line proof
彩色样图 colour manuscript
彩色增强 colour enhancement
彩色坐标系 colour coordinate system
参考高度 reference altitude
参考椭球 reference ellipsoid
参考椭球定位 orientation of reference ellipsoid
参考系 frame of reference
参考值 reference value
参量效应 parametric effect

参数平差 parameter adjustment
参照效应 reference effect
糙率 roughness
槽波地震法 inseam seismic method
槽式结构 trough-like structure
草地 grassland
草甸 meadow
侧方交会 side intersection
侧滚角 angle of roll
侧击波 side swipe
侧面波 lateral wave
侧面的 lateral
侧扫声呐 side scan sonar
侧视雷达 side-looking radar
侧向测井 laterolog
测标 mark
测地子午线 geodetic meridian
测点 survey station
测杆 measuring bar
测高 altimetry
测高法 altimetry
测高仪 alidade; altimeter
测轨系统 trajectory measuring system
测绘标准 standard of surveying and mapping
测绘工程 surveying and mapping engineering
测绘学 geomatics; SM (surveying and mapping)
测绘仪器 instrument of surveying and mapping
测角中误差 mean square error of angle observation
测井 logging; well logging
测井技术 logging technology

测井曲线 logging curve
测井数据 logging data
测井图 log
测井仪 logging tool
测距 ranging
测距定位系统 range positioning system
测距雷达 range-only radar
测距盲区 range hole
测距仪 rangefinder
测控条 control strip
测量 measurement
测量标志 survey mark
测量船 survey vessel
测量电极 potential electrode
测量规范 specification of surveys
测量精度 measurement precision
测量控制网 surveying control network
测量平差 adjustment of observation; survey adjustment
测量数据 measured data
测量误差 measurement error
测量学 surveying
测流 current surveying
测深 sounding
测深法 sounding
测深改正 correction of depth
测深杆 sounding pole
测深精度 total accuracy of sounding
测深仪读数精度 reading accuracy of sounder
测深仪发射线 transmitting line of sounder
测深仪回波信号 echo signal of sounder
测深仪记录纸 recording paper of sounder
测深仪零线 transmitting line of sounder
测速标 mark for measuring velocity
测图 mapping
测图卫星 mapping satellite

测网 survey grid
测微密度计 microdensitometer
测微目镜 micrometer eyepiece
测微器 micrometer
测线 survey line
测线束法 swath
测站 station; survey station
测站归心 station centring
测震学 seismometry
层间层 intervening layer
层间改正 plate correction
层面改正 bedding correction
层速度 interval velocity
层位 horizon
层位拉平 horizon flattening
层析 tomographic
层析成像 tomography
层析地震成像 seismic tomography
层状畴 lamellar domain
层状的 layered
差分台 track station
差频 beat frequency
差异阈 difference threshold
插值 interpolation
查普曼层 Chapman layer
查普曼生成函数 Chapman production function
查询 query
觇牌 target
掺气 aeration
产额函数 yield function
产流 runoff yield
产沙(量) sediment yield
长波辐射 long-wave radiation
长度标准检定场 standard field of length
长期变化 long-term variation; secular change; secular variation
长期均值 long-term average
长周期循环 long-period circulation
常绿阔叶林 evergreen broad-leaved forest
常相位关系 constant phase relationship

厂址测量 surveying for site selection
场差 field difference
场地烈度 site intensity
场反向 field-reversal
场分布 field distribution
场向不规则结构 field-aligned irregularity
场向电流 field-aligned current
场形变 field deformation
场致反向 field-reversal
超变质水 ultrametamorphic water
超导磁力仪 SQUID magnetometer (superconductive magnetometer)
超导性 superconductivity
超导重力仪 superconductive gravimeter
超高频 ultra-high frequency
超焦点距离 hyperfocal distance
超近摄影测量 macrophotogrammetry
超平流层 super stratosphere
超日冕 super-corona
超声成像测井 ultrasonic image logging
超声极风 supersonic polar wind
超声膨胀 supersonic expansion
超声速太阳风 supersonic solar wind
超旋转的 superrotational
超压 overpressure
潮汐表 tidal table
潮汐波 tidal wave
潮汐带地震勘探 tidal zone seismic
潮汐的 tidal
潮汐非调和常数 tidal nonharmonic constant
潮汐非调和分析 tidal nonharmonic analysis
潮汐摄动 tidal perturbation
潮汐调和常数 tidal harmonic constant
潮汐调和分析 tidal harmonic analysis
潮汐效应 tidal effect
潮汐因子 tidal factor
潮汐预报 tidal prediction
潮汐运动 tidal motion
潮汐振荡 tidal oscillation
潮汐作用 tidal action
潮信表 tidal information panel
沉积后碎屑剩磁 post-depositional DRM
沉积剩磁 depositional remanence; DRM (depositional remanent magnetization)
沉积碎屑剩磁 depositional DRM
沉积物地球化学 sedimentary geochemistry
沉降 settling
沉降电子 precipitating electron
沉降观测 settlement observation
沉降通量 precipitation flux
晨昏电场 dawn-dusk electric field
晨昏电场〔增强的〕 increased dawn-dusk electric field
晨昏蒙影 twilight
晨昏子午线 dawn and dusk meridian
成矿地球化学异常 ore geochemical anomaly
成像 imaging
成像光谱仪 imaging spectrometer
成像雷达 imaging radar
城市测量 urban survey
城市地形测量 urban topographic survey
城市地形图 topographic map of urban area
城市基础地理信息系统 UGIS (urban geographical information system)
城市景观 urban landscape
城市控制测量 urban control survey
城市土地 urban land

城市制图 urban mapping
乘常数 multiplication constant
弛豫时间 relaxation time
尺度参数 scale parameter
尺度定律 scaling law
赤道 equator
赤道半径 equatorial radius
赤道的 equatorial
赤道电集流 equatorial electrojet
赤道电离层 equatorial ionosphere
赤道卫星 equatorial satellite
赤道吸收 equatorial absorption
赤道异常 equatorial anomaly
赤经 right ascension
赤纬圈 declination circle
冲击带 impact zone
冲击加速度 impact acceleration
冲积的 alluvial
冲角 angle of attack
冲刷 scour
充电法 excitation-at-the-mass method; mise-a-la-masse method
重联 reconnection
重联速率 reconnection rate
重现频率 recurrence frequency
重现周期 recurrent period; return period
抽象符号 abstract symbol
臭氧层 ozonosphere
臭氧层顶 ozonopause
臭氧垂直剖面 vertical ozone profile
臭氧反应 ozone reaction
臭氧分布 ozone distribution
臭氧分光光度计 ozone spectrophotometer
臭氧光解 ozone photolysis
臭氧截断 ozone cutoff
臭氧量 ozone amount
臭氧浓度 ozone concentration
臭氧吸收 ozone absorption
臭氧形成 ozone formation
出版原图 final original
初定震中 PDE (Preliminary Determination of Epicentre)
初动 first motion; first movement
初动近似 first motion approximation
初生大气 primordial atmosphere
初始参考地球模型 PREM (Preliminary Reference Earth Model)
初始扰动 initial disturbance
初始应力 initial stress; virgin stress
初相 initial phase
初至波 primary wave
储层地震地层学 reservoir seismic stratigraphy
储集层 reservoir
触地点 earth point
触发 triggering
触发机制 trigger mechanism
触发效应 trigger effect
触发作用 trigger action
触觉地图 tactual map
穿透辐射 penetrating radiation
穿透深度 penetration depth
传播 propagation
传播矩阵 propagator matrix
传播延迟 propagation delay
传导热流 conductive heat flow
传递函数 transfer function
船台 mobile station
船载重力仪 shipboard gravimeter
垂核面 vertical epipolar plane
垂核线 vertical epipolar line
垂球 plumb bob
垂线偏差 deflection of the vertical
垂线偏差改正 correction for deflection of the vertical
垂直不稳定性 vertical instability
垂直地震测线法 VSP survey (vertical seismic profiles survey)
垂直地震剖面 vertical seismic profile

垂直地震剖面测井 vertical seismic profile log
垂直叠加 vertical stacking
垂直分布 vertical distribution
垂直分量 vertical component
垂直共面线圈系统 vertical coplanar coils system
垂直光学厚度 vertical optical thickness
垂直截面 vertical section
垂直密度剖面 vertical density profile
垂直能量通量 vertical energy flux
垂直漂移 vertical drift
垂直气压梯度 vertical pressure gradient
垂直强度 vertical intensity
垂直切变 vertical shear
垂直切变矢量 vertical shear vector
垂直圈 vertical circle
垂直入射 vertical incidence
垂直输送 vertical transport
垂直通量 vertical flux
垂直同轴线圈系统 vertical coaxial coils system
垂直折光误差 vertical refraction error
垂直折光系数 vertical refraction coefficient
垂准仪 plumb aligner
春分点 First Point of Aries
春季极大值 spring maximum
纯声波 pure sound wave
纯重力异常 pure gravity anomaly
磁暴 geomagnetic storm; magnetic storm
磁暴后效 after-effect of (magnetic) storm
磁暴急始 storm sudden commencement
磁暴形态学 storm morphology
磁北极 north magnetic pole
磁泵过程 magnetic pumping process
磁变年差 annual change of magnetic variation
磁变仪 variometer
磁测 geomagnetic survey
磁测深 magnetic sounding
磁测深线 magnetic sounder
磁测深仪 magnetic sounder
磁层 magnetosphere
磁层暴 magnetospheric storm
磁层等离子体 magnetospheric plasma
磁层顶 magnetopause
磁层对流 magnetospheric convection
磁层倾泻 magnetospheric dumping
磁层热等离子体 hot magnetospheric plasma
磁层响应 magnetospheric response
磁层状态 state of the magnetosphere
磁场 magnetic field
磁场变化 magnetic field change
磁场极性 magnetic field polarity
磁场能量湮没 annihilation of magnetic field energy
磁场梯度漂移 magnetic gradient drift
磁充电法 magnetic charging method
磁大地电流法 magnetotelluric method
磁地方时 magnetic local time
磁叠印 magnetic overprinting
磁法调查 magnetic survey
磁法勘探 magnetic prospecting
磁方位角 magnetic azimuth
磁刚度 magnetic rigidity
磁共轭点 magnetic conjugate point
磁钩扰 magnetic crochet
磁化方向 direction of magnetization
磁化率 susceptibility

磁化率测井 magnetic susceptibility logging
磁化率计 magnetic susceptibility meter
磁激发极化法 MIP method (magnetic induced polarization method)
磁极归化 reduced to the magnetic pole
磁晶各向异性 magnetocrystalline anisotropy
磁静带 magnetic quiet zone
磁静日 magnetically quiet day
磁矩 magnetic moment
磁壳 magnetic shell
磁离子理论 magneto-ionic theory
磁力扫海测量 magnetic sweeping
磁力梯度仪 magnetic gradiometer
磁力线图 magnetic figure
磁力仪 magnetometer
磁力异常区 magnetic anomaly area
磁流体力学 hydromagnetics
磁南极 south magnetic pole
磁能湮没 annihilation of magnetic field energy
磁偶极 magnetic dipole
磁偶极时 magnetic dipole time
磁耦合 magnetic coupling
磁偏角 declination; grid variation; magnetic declination; magnetic variation
磁谱仪 magnetic spectrograph
磁鞘 magnetosheath
磁鞘区 magnetosheath region
磁倾极 dip pole
磁倾计 inclinometer
磁倾角 dip angle; inclination; magnetic dip; magnetic dip angle
磁倾圈 dip circle
磁清洗 magnetic cleaning; magnetic washing
磁情记数 magnetic character figure
磁扰 magnetic disturbance
磁扰活动性 geomagnetic disturbance activity
磁扰日 magnetically disturbed day
磁扇形结构 magnetic sector structure
磁通计 fluxmeter
磁通量计 fluxmeter
磁通量密度 magnetic flux density
磁通门磁力仪 flux-gate magnetometer
磁图 magnetic chart
磁湾扰 magnetic bay
磁尾 magnetotail
磁尾张力 tension in the tail
磁位形 magnetic configuration
磁象限角 magnetic bearing
磁效应 magnetic effect
磁性地层学 magnetic stratigraphy; magnetostratigraphy
磁压 magnetic pressure
磁亚暴 magnetic substorm
磁异常特征 magnetic signature
磁余纬 magnetic colatitude
磁元 magnetic element
磁源重力异常 gravity anomaly due to magnetic body; pseudogravity anomaly
磁照图 magnetogram
磁子午线 magnetic meridian
磁组构 magnetic fabric
磁坐标 magnetic coordinate
次级电子 secondary electron
次级辐射 secondary radiation
次级离子 secondary ion
次生磁化(强度) secondary magnetization
次生地球化学异常 secondary geochemical anomaly
次生剩磁 secondary remanent magnetization

粗差 gross error
粗差检测 gross error detection
粗码 C/A Code (Coarse/Acquision Code)

淬灭反应 quenching reaction
淬灭剂 quenching agent
措尔纳悬挂法 Zollner suspension

打样 proofing
大扁度轨道 highly elongated orbit
大尺度辐合 large-scale convergence
大尺度太阳磁场 large-scale solar magnetic field
大尺度行星波 large-scale planetary wave
大地测量 geodetic surveying
大地测量的 geodetic
大地电磁测深 magnetotelluric sounding
大地电磁的 magnetotelluric
大地电流法 telluric (current) method
大地方位角 geodetic azimuth
大地高 ellipsoidal height; geodetic height
大地基准 geodetic datum
大地经度 geodetic longitude
大地热流 terrestrial heat flow
大地水准面 geoid
大地天顶延迟 atmosphere zenith delay
大地天文学 geodetic astronomy
大地纬度 geodetic latitude
大地位 geopotential
大地线 geodesic
大地线微分方程 differential equation of geodesic
大地重力学 physical geodesy
大地主题反解 inverse solution of geodetic problem
大地主题正解 direct solution of geodetic problem
大地坐标 geodetic coordinate
大地坐标系 geodetic coordinate system
大旱 severe drought
大红斑 Great Red Spot
大环投影 gnomonic projection
大角度反射 wide-angle reflection
大孔径地震台阵 LASA (large-aperture seismic array)
大陆板块 continental plate
大陆重建 continental reconstruction
大陆分裂 continental splitting
大陆扩张 continental spreading
大陆漂移 continental drift
大陆拼合 continental fitting
大气边界 atmospheric boundary
大气边缘层 fringe region of atmosphere
大气标高 atmosphere scale height
大气不透明度 atmospheric opacity
大气参数 atmospheric parameter
大气潮汐 atmospheric tide
大气窗 atmospheric window
大气窗区 atmospheric window region
大气簇射 air shower
大气地球化学 geochemistry of the atmosphere
大气电离 atmospheric ionization
大气电学 atmospheric electricity
大气发射 atmospheric emission
大气辐射 atmospheric radiation
大气光学厚度 atmosphere optical thickness

大气光学质量 optical air mass
大气过程 atmospheric process
大气环流 atmospheric circulation
大气浑浊度 atmospheric opacity; atmospheric turbidity
大气结构 atmospheric structure
大气扩散 atmospheric diffusion
大气离子 atmospheric ion
大气密度 atmospheric density
大气模式 atmospheric model
大气谱带 atmospheric band
大气水 meteoric water
大气湍流 atmospheric turbulence
大气涡度 atmospheric vorticity
大气污染 air pollution
大气吸收 atmospheric absorption
大气响应 atmospheric response
大气消光 atmospheric extinction
大气压 atmospheric pressure
大气余辉发射 air afterglow emission
大气折射 atmospheric refraction
大气振荡 atmospheric oscillation
大气制动 atmospheric braking
大扰动 major disturbance
大行星 major planet
大振幅理论 large amplitude theory
大震 major earthquake
大震速报台网 Large Earthquake Prompt Report Network
带宽 band width
带状地形图 belt topographic map
单侧断裂 unilateral faulting
单畴颗粒 single domain particle
单点 single point
单峰分布 unimodal distribution
单个带电粒子 solitary charged particle
单级火箭 single-stage rocket
单极磁区 unipolar magnetic region
单极黑子 unipolar sunspot
单极群 unipolar group
单井 single well
单色的 monochromatic
单色光度计 filter photometer
单向通量 unidirectional flux
单源单缆海上地震采集 offshore single-source and single-streamer seismic acquisition
弹道学 ballistics
氮循环 nitrogen cycle
氮原子 nitrogen atom
当地子午线 local meridian
氘 deuterium
氘核 deuteron
导管 duct
导管传播 ducted propagation
导航 navigation
导航定位 navigation positioning
导航系统 navigation system
导线 traverse
导炸索 explosive cord
岛弧地热带 island arc geothermal zone
到达角 angle of arrival
到时 arrival time
到时差 arrival time difference
倒 V 事件 inverted-V event
倒转 retrograde
倒转检验 reversal test
倒转时间 time of reversal
道间均衡 trace equalization
道内动平衡 dynamic equalization
德国官方地形制图信息系统 Authoritative Topographic Cartographic Information System
德胡普变换 De Hoop transformation
德洛内三角网 Delaunay triangulation
等变形线 distortion isogram
等磁强线 isomagnetic line
等磁倾线 magnetic isoclinic line

等磁图 isomagnetic chart
等磁异常线 magnetic isoanomalous line
等地温面 geoisotherm; isogeotherm
等地温线 isogeotherm
等高圈 almucantar
等高线 contour
等积投影 equivalent projection
等级感 ordered perception
等级结构 hierarchical organization
等价性原理 principle of equivalence
等角投影 conformal projection
等距量表 interval scaling
等距投影 equidistant projection
等离子体不稳定性 plasma instability
等离子体参考系 plasma frame of reference
等离子体层 plasmasphere
等离子体层顶 plasmapause
等离子体共振 plasma resonance
等离子体加热 heating of the plasma
等离子体幔 plasma mantle
等离子体抛射 plasma ejection
等离子体片 plasmasheet
等离子体频率 plasma frequency
等离子体物理学 plasma physics
等量纬度 isometric latitude
等年变线 isopore; isoporic line
等深流丘状地震相 contourite mound seismic facies
等体积波 equivoluminal wave
等位线 equipotential line
等温层 isothermal layer
等温剩磁 IRM (isothermal remanent magnetization)
等吸收线 isoabsorption line
等效电流系 equivalent current system
等效空气程 equivalent air path
等压线 isopiestics
等震线 isoseismal curve; isoseismal line
等值灰度尺 equal value gray scale
等值区域图 choroplethic map
等值线 contour line
等值线地图 isoline map
等值线法 isoline method
等值线图 contour map
等值线图偏移 contour map migration
低层大气 lower atmosphere
低电离层 lower ionosphere
低轨卫星 low-orbiting satellite
低空火箭 low-altitude rocket vehicle
低能电子 low-energy electron
低能粒子 low-energy particle
低热层 lower thermosphere
低水位 low water
低速层 LVL (low velocity layer)
低速区 LVZ (low velocity zone)
低纬磁暴 low-latitude storm
低纬极光 low-latitude aurora
低氧丰度 low oxygen abundance
笛卡儿坐标 Cartesian coordinate
底部电离层 bottom-side ionosphere
底部频高图 bottom-side ionogram
底点纬度 latitude of pedal
底色去除 under colour removal
底色增益 under colour addition
底视探测仪 bottom-side sounder
地扁球体 earth spheroid
地表 earth surface
地表波 ground wave
地表地球化学测量 surface geochemical survey
地表地热显示 surface geothermal manifestation
地表热流 surface heat flow
地层 formation
地层压力 formation pressure

地磁 geomagnetism
地磁测量 geomagnetic survey
地磁场 geomagnetic field
地磁赤道 geomagnetic equator
地磁的 geomagnetic
地磁观测 geomagnetic observation
地磁活动 geomagnetic activity
地磁极 geomagnetic pole
地磁极性反向 geomagnetic polarity reversal
地磁极性年表 geomagnetic polarity time scale
地磁极性转向年表 geomagnetic polarity reversal time scale
地磁静止动量 geomagnetic cut-off momentum
地磁年代学 geomagnetic chronology
地磁漂移 geomagnetic excursion
地磁台 geomagnetic station
地磁微脉动 magnetic micropulsation
地磁物理学 geomagnetic physics
地磁响应 geomagnetic response
地磁学 geomagnetism
地磁要素 magnetic element
地磁指数 geomagnetic index
地磁轴 geomagnetic axis
地磁坐标 geomagnetic coordinate
地磁坐标系 geomagnetic coordinate system
地电的 geoelectric
地电断面 geoelectric cross section
地电阻率 earth resistivity
地方恒星时 local sidereal time
地方时分布 local time distribution
地方视时 local apparent time
地方视正午 local apparent noon
地方震 local earthquake; local shock
地方震级 local magnitude
地滚 ground roll
地核磁场 core field

地籍 cadastre
地籍测量 cadastral survey; cadastration
地籍制图 cadastral mapping
地价 land price
地理格网 geographic grid
地理空间 geographical space
地理信息 geographic information
地理信息传输 geographic information communication
地理信息系统技术 GIS technology
地理学 geography
地理坐标 geographic coordinate
地脉动 earth pulsation
地幔 (earth) mantle
地幔对流 mantle convection
地幔对流环 mantle convection cell
地幔热流 mantle heat flow
地幔焰 mantle plume
地幔柱 mantle plume
地貌 landform; topographic feature
地貌调查 topographic feature survey
地貌图 geomorphological map
地貌形态示量图 morphometric map
地貌学 geomorphology
地冕 geocorona
地面测量 ground-based measurement
地面查证 ground follow-up
地面沉降 ground subsidence
地面反向散射 ground-backscatter
地面-井中方式 surface-borehole variant
地面气压 surface pressure
地面纬圈 terrestrial parallel
地面运动 ground motion
地平经度 azimuth
地平经圈 azimuth circle; vertical circle

地平经纬仪 altazimuth
地平纬圈 almucantar
地平坐标系 horizontal system of coordinate
地壳 crust
地壳传递函数 crustal transfer function
地壳的 crustal
地壳地震 crustal earthquake
地壳构造 earth crust structure
地壳均衡(说) isostasy
地壳形变 crustal deformation
地倾斜 earth tilt; ground tilt
地球 earth
地球变平换算 earth-flattening transformation
地球变平近似 earth-flattening approximation
地球动力学 geodynamics
地球反照率 albedo of the Earth
地球辐射 terrestrial radiation
地球辐射通量 terrestrial radiation flux
地球干涉量度学 terrestrial interferometry
地球化学 geochemistry
地球化学标准元素 geochemical leading element
地球化学地方病 geochemical endemic
地球化学地势 geochemical relief
地球化学分类 geochemical classification
地球化学分散 geochemical dispersion
地球化学分异酌 geochemical differentiation
地球化学过程 geochemical process
地球化学环境 geochemical environment
地球化学基本定律 basic geochemical law
地球化学景观 geochemical landscape
地球化学景观〖从属的〗 subordinate geochemical landscape
地球化学勘探 geochemical exploration; geochemical prospecting; geochemical survey
地球化学面 geochemical surface
地球化学平衡 geochemical balance
地球化学剖面 geochemical profile
地球化学区 geochemical province
地球化学普查 geochemical recognition
地球化学生态学 geochemical ecology
地球化学水系普查 geochemical drainage reconnaissance
地球化学踏勘 geochemical reconnaissance
地球化学探矿 geochemical prospecting
地球化学梯度 geochemical gradient
地球化学填图 geochemical mapping
地球化学图 geochemical map
地球化学详查 geochemical detailed survey
地球化学相 geochemical facies
地球化学性质〖元素的〗 geochemical character of elements
地球化学性状 geochemical behaviour
地球化学蓄电池假说 hypothesis of geochemical accumulators
地球化学异常 geochemical anomaly
地球化学障 geochemical barrier
地球化学者 geochemical province
地球化学指标 geochemical indicator

地球化学指数 geochemical index
地球纪年学 geochronology
地球模型 earth model
地球谱学 terrestrial spectroscopy
地球事件 terrestrial event
地球椭率 compression of the earth
地球纬度圈 terrestrial parallel
地球位数 geopotential number
地球温度计 geothermometer
地球物理测井 geophysical well-logging
地球物理场 geophysical field
地球物理的 geophysical
地球物理方法 geophysical method
地球物理勘探 geophysical exploration; geophysical prospecting
地球物理学 geophysics
地球物理异常 geophysical anomaly
地球向阳侧 sunward side of the earth
地球效应 terrestrial effect
地球形状 figure of the earth
地球仪 globe
地球子午线 terrestrial meridian
地球自转 earth rotation
地热 geoheat; geotherm
地热的 geothermal
地热调查 geothermal survey
地热活动 geothermal activity
地热勘探 geothermal prospecting
地热流体 geofluid; geothermal fluid
地热能 geothermal energy
地热水库 geothermal reservoir
地热田 geothermal field
地热系统 geothermal system
地热现象 geothermal phenomenon
地热学 geothermics
地热异常 geothermal anomaly
地热异常区 geothermally-anomalous area
地热资源 geothermal resources
地声 earthquake sound
地势图 hypsometric map
地图编辑 map editing
地图编辑大纲 map editorial policy
地图编制 map compilation
地图表示法 cartographic presentation
地图传输 cadographic communication; cartographic communication
地图叠置分析 map overlay analysis
地图分类 cadographic classification; cartographic classification
地图分析 cadographic analysis; cartographic analysis
地图符号库 map symbols bank
地图符号学 cartographic semiology
地图负载量 map load
地图复杂性 map complexity
地图复制 map reproduction
地图感受 map perception
地图更新 map revision
地图集 atlas
地图集信息系统 atlas information system
地图利用 map use
地图量算法 cartometry
地图模型 cartographic model
地图内容结构 cartographic organization
地图判读 map interpretation
地图评价 cartographic evaluation
地图潜信息 cartographic potential information
地图清晰性 map clarity
地图色标 map colour standard
地图色谱 map colour atlas

地图设计　map design
地图数据结构　map data structure
地图数字化　map digitizing
地图投影　map projection
地图显示　map display
地图信息　cadographic information; cartographic information
地图信息系统　CIS (cartographic information system)
地图学　cartography
地图研究法　cartographic methodology
地图易读性　map legibility
地图印刷　map printing
地图语法　cartographic syntactics
地图语言　cartographic language
地图语义　cartographic semantics
地图语用　cartographic pragmatics
地图整饰　map decoration
地图制图　cartography; map making
地外源　extra-terrestrial origin
地外震学　extra-terrestrial seismology
地温梯度　geothermal gradient
地下流体　underground fluid
地下热水　geothermal water
地下水　groundwater
地下水位　groundwater level
地下网格　subsurface grid
地心的　geocentric
地心坐标　geocentric coordinate
地形　terrain; topography
地形等压线　topographic isobar
地形校正　terrain correction; topographic correction
地形控制点　topographic control point
地形罗斯贝波　topographic Rossby wave
地形数据库　topographic database
地形同位素分馏效应　topographic isotopic fractionation effect
地形图　topographic map
地学　geoscience
地月空间　cislunar space
地震　earthquake
地震标准层　seismic marker horizon
地震波　earthquake wave; seismic wave
地震波频散　seismic-wave dispersion
地震波识别　identification of seismic events
地震波走时　travel time
地震参数　seismic parameter
地震测井　well shooting
地震测深　seismic sounding
地震层位　seismic horizon
地震成因　cause of earthquake
地震重复率　earthquake recurrence rate
地震储层研究　seismic reservoir study
地震触发器　seismic trigger
地震大小　shock size
地震带　seismic belt
地震道　seismic channel
地震等浮电缆　seismic streamer
地震地质学　seismic geology; seismogeology
地震调查　seismic survey
地震定位　earthquake location
地震动　seismic ground motion
地震动土压力　seismic earth pressure
地震断层　seismic fault
地震反射　seismic reflection
地震反射法　seismic reflection method
地震反演　seismic inversion
地震工程（学）　earthquake engineering
地震构造带　seismic-tectonic zone
地震构造区　seismotectonic province

地震构造图 seismic structural map
地震构造学 seismic tectonics; seismotectonics
地震光 earthquake light
地震海啸 seismic sea wave
地震活动带 seismically active belt
地震活动区 seismically active zone
地震活动性 seismic activity; seismicity
地震活动性图像 seismicity pattern
地震机制 earthquake mechanism
地震计 seismometer
地震记录 seismic record
地震记录仪 seismic recording instrument
地震技术 seismic technology
地震加速度 seismic acceleration
地震监测 seismic surveillance
地震警报 earthquake warning
地震矩 seismic moment
地震勘探 seismic exploration; seismic prospecting
地震勘探暗点 dim spot
地震勘探船 seismic survey vessel
地震勘探等浮电缆 seismic streamer
地震勘探亮点 bright spot
地震空区 seismic gap
地震力 earthquake force
地震量 seismic mass
地震烈度 earthquake intensity; seismic intensity
地震烈度复核 checkup of seismic intensity
地震轮回 seismic cycle
地震面波 seismic surface wave
地震模型 seismic model; seismology model
地震模型学 seismology model
地震目录 earthquake catalogue
地震能量 seismic energy
地震频度 earthquake frequency

地震破裂 seismic rupture
地震破裂力学 earthquake rupture mechanics
地震剖面 seismic section
地震迁移 earthquake migration
地震区 earthquake province; earthquake region; seismic zone
地震区划 seismic regionalization; seismic zoning
地震社会学 seismosociology
地震射线 seismic ray
地震数据预处理 seismic data preprocessing
地震速度 seismic velocity
地震台 seismic station
地震台精密测量 precise survey at seismic station
地震台网 seismic network
地震体波 bodily seismic wave; seismic body wave
地震统计(学) earthquake statistics
地震图 seismogram
地震危险区 earthquake-prone area
地震危险性 earthquake risk; seismic risk
地震危险性分析 seismic risk analysis
地震危险性评定 seismic risk evaluation
地震位错 earthquake dislocation; seismic dislocation
地震物探法 seismic geophysical survey
地震吸收带 seismic absorption band
地震系列 earthquake series
地震系数法 seismic coefficient method
地震响应 seismic response
地震响应分析 analysis of structural response to seismic excitation
地震相 seismic facies
地震相单元 seismic facies unit

地震相分析 seismic facies analysis
地震相勘探 seismic facies
地震相图 seismic facies map
地震效率 seismic efficiency
地震信号 seismic signal
地震序列 earthquake sequence; seismic sequence
地震学 seismology
地震研究观测台 SRO (Seismic Research Observatory)
地震仪 seismic recording instrument; seismograph
地震预报 earthquake forecasting
地震预测 earthquake prediction
地震预防 earthquake prevention
地震孕育 earthquake preparation
地震灾害 earthquake disaster
地震载荷 earthquake loading
地震站 seismic station
地震折射法 seismic refraction method
地震震级 magnitude
地震周期 earthquake period
地震周期性 earthquake periodicity
地震子波 seismic wavelet
地质的 geological
地质雷达 geological radar; ground penetrating radar
地质年代学 geochronology
地质平面图 geological map
地质图 geological map
第三不变量 third invariant
第三期蠕变 tertiary creep
第三宇宙速度 third cosmic velocity
第四纪 quaternary
第一宇宙速度 first cosmic velocity
缔合脱离 associative detachment
典型反应时间 typical reaction time
点方式 point mode
点位 point position

点值法 dot method
点状符号 point symbol
电测井 electrical logging
电测深 electrical sounding
电磁波 electromagnetic wave
电磁测深 electromagnetic sounding
电磁场 electromagnetic field
电磁的 electromagnetic
电磁法 electromagnetic method
电磁辐射 electromagnetic radiation
电磁感应法 electromagnetic induction method
电磁脉冲震源 electromagnetic vibration exciter
电磁式地震仪 electromagnetic seismograph
电导率 electric conductivity
电导率测井 conductivity logging
电导率张量 conductivity tensor
电动力效应 electro-dynamic effect
电法调查 electrical survey
电法勘探 electrical prospecting
电荷交换反应 charge-exchange reaction
电弧放电 arc discharge
电花光谱 spark spectrum
电极电位测井 electrode potential logging
电极化场 electric polarization field
电极排列 electrode array
电集流 electrojet
电离层 ionosphere
电离层D区 ionospheric D-region
电离层暴 ionospheric storm
电离层测高仪 ionosonde
电离层的 ionospheric
电离层等离子体 ionospheric plasma
电离层顶 ionopause
电离层改造 ionospheric modification
电离层剖面 ionospheric profile

电离层扰动 ionospheric disturbance
电离层食 ionospheric eclipse
电离层吸收 ionospheric absorption
电离层行扰 TID (travelling ionospheric disturbance)
电离大气 ionized atmosphere
电离度 degree of ionization
电离辐射 ionizing radiation
电离率 ionization rate
电离密度 density of ionization
电离能 ionization energy
电离区 ionized region
电离突增 sudden increase of ionization
电离图 ionogram
电位 potential
电位梯度 potential gradient
电文学 ionospherics
电晕放电 corona discharge
电子出版系统 electronic publishing system
电子地图 electronic map
电子地图集 electronic atlas
电子感应加速器 betatron
电子感应加速效应 betatron effect
电子化学 electron chemistry
电子回旋加速器 betatron
电子极光 electron aurora
电子极光带 electron zone
电子径迹 track of electron
电子密度 electron density
电子能谱 electron energy spectrum
电子总含量 TEC (total electron content)
电阻率 resistivity
电阻率测井 resistivity logging
电阻率法 resistivity method
电阻率剖面法 resistivity profiling
电阻性元件 resistive element
凋落物 litter
吊舱系统 towed bird system

叠合解释 overlay interpretation
叠后偏移 post stack migration
叠加 overlay; stacking
叠加平均 overlapping average
叠加剖面图 stacked profiles map
叠加速度 stacking velocity
叠前偏移 prestack migration
叠印 overprint
蝶形图 butterfly diagram
顶视探测仪 topside sounder
顶外电离层 topside ionosphere
定常轨道 stationary orbit
定量分析 quantitative analysis
定位 positioning
定位技术 positioning technology
定位检索 retrieval by window
定位精度 positioning precision
定位统计图表法 positioning diagram method
定位系统 positioning system
定向 orientation
定向天线 beam antenna
定向运动地图 orienteering map
定性分析 qualitative analysis
定性检索 retrieval by header
定源场 fixed source field
定源法 fixed source method
东西效应 asymmetric effect
东西星等高测时法 method of time determination by Zinger star-pair
冬半球 winter hemisphere
冬季异常 winter anomaly
冬至 winter solstice
氡气测量 radon survey
动感 autokinetic effect
动画引导 animated steering
动画制图 animated mapping
动校正 NMO correction (normal moveout correction)
动力大地测量学 dynamic geodesy
动力太阳风 dynamic solar wind

动能密度 kinetic energy density
动能转换 kinetic energy transformation
动生感应 motional induction
动态变量 dynamic variable
动态地景仿真 dynamic landscape simulation
动态地图 dynamic map
动态范围 dynamic range
动态机械放大倍数 dynamical mechanical magnification
动源法 moving source method
冻结核 freezing nucleus
洞穴 cave
陡峭入射 steep incidence
渡越路径 transit path
短棒图 stick plot
短波辐射 short wave radiation
短波频率急偏 SFD (sudden frequency deviation)
短波衰退 SWF (short wave fade-out)
短波通讯中断 blackout; fade-out
短波突然衰退 sudden short wave fade-out
短路 short circuits
短期响应 short-term response
短期涨落 short-term fluctuation
短周期彗星 short-period comet
短轴 minor axis
断层 fault; faulting
断层地震 fault earthquake
断层面解 fault-plane solution
断层作用 faulting
断块 fault block
断裂 fracture; fault
断裂带 fault zone
断面 section
断面图 section map
断陷 fault depression
断陷盆地 fault basin
对称剖面法 symmetrical profiling
对称四极测深 symmetrical four-pole sounding
对称振型 symmetrical mode
对地静止卫星 fixed satellite
对流 convection
对流层 troposphere
对流层顶 tropopause
对流层、平流层、中层大气探测雷达 MST radar
对流层事件 tropospheric event
对流环 convection cell
对流区 convective region
对流热流 convective heat flow
对流图型 convection pattern
对日照 Gegenschein
对准 registration
钝角 obtuse angle
多边形 polygon
多边形地图 polygonal map
多边形结构 polygon structure
多波 multiwave
多层结构 multi layer organization
多层球指数 polytropic index
多重地震 multiple earthquake
多畴颗粒 multidomain grain
多畴热剩磁 multidomain thermal remanence
多次反射 multiple reflection
多次覆盖 multiple coverage
多次散射 multiple scattering
多道地震仪 multichannel seismic instrument
多焦点投影 polyfocal projection
多路编排 multiplex
多路解编 demultiplex
多媒体地图 multimedia map
多频振幅相位法 multiple frequency amplitude-phase method
多普勒频移 Doppler frequency shift
多项式 polynomial
多圆锥投影 polyconic projection
惰性气体 inert gas

俄歇簇射 Auger shower
厄缶改正 Eotvos correction
厄特沃什改正 Eotvos correction
蒽 anthracene
二乘 square
二分点 equinoctial point
二进制码 binary code
二维偶极模型 two-dimensional dipole model

二氧化氮 nitrogen dioxide
二氧化氮反应 nitrogen dioxide reaction
二氧化氮光化学 nitrogen dioxide photochemistry
二氧化碳 carbon dioxide
二元等离子体 two-component plasma

发电机区　dynamo region
发光谱　luminescence spectrum
发光区　luminous region
发光效率　luminous efficiency
发光云实验　glow cloud experiment
发散带　divergence belt; divergence zone
发射法　shooting method
发射方位角　launch azimuth
发射谱线　emission line
发射系数　emission coefficient
发生频次　occurrence frequency
发声火流星　detonating fireball; sound-emitting fireball
发震时刻　origin time
发震应力　earthquake-generating stress
法截面　normal section
法拉第效应　Faraday effect
法拉第旋转　Faraday rotation
法律地震学　forensic seismology
法向分量　normal component
反常彗尾　anomalous tail
反常色散　anomalous dispersion
反对称振型　antisymmetrical mode
反符合电路　anticoincidence circuit
反频散　inverse dispersion
反平面剪切裂纹　anti-plane shear crack
反气旋　anticyclone
反气旋脊　anticyclonic ridge
反山根　antiroot
反射　reflection
反射波　reflection wave
反射地震学　reflection seismology
反射电子　mirror electron
反射矩阵　reflection matrix
反射率法　reflectivity method
反射频率　reflected frequency
反射系数　reflection coefficient
反射效应　reflecting effect
反向极性　reversed polarity
反像　mirror reverse
反演　inversion
反应链　reaction chain
反应速率系数　reaction rate coefficient
反照电子　albedo electron
反照率　albedo
反照中子　albedo neutron
反照中子理论　albedo neutron theory
反照中子能谱　albedo neutron spectrum
反褶积　deconvolution
范艾仑带　Van Allen belt
范艾仑辐射带　Van Allen belt
范围法　area method
方里网　kilometer grid
方位　azimuth; orientation
方位对称性　azimuthal symmetry
方位角　azimuth
方位角分量　azimuthal component
方位距离定位方法　azimuth distance positioning method; polar positioning method
方位漂移　azimuthal drift
方位圈　azimuth circle
方位投影　azimuthal projection

方向观测法 method of direction observation
方向性 directivity
方向性函数 directivity function
方向余弦 direction cosine
防灾 disaster prevention
防震缝 aseismic joint
仿射的 affine
仿射绘图 affine plotting
放射计 radiometer
放射性测井 radioactivity logging
放射性沉降物 radioactive fallout
放射性调查 radioactivity survey
放射性勘探 radioactivity prospecting
放射性示踪测井 radioactive tracer logging
放射性衰变 radioactive decay
放射性碳 radiocarbon
放射医学 radiation medicine
飞行特性 flight characteristic
非磁化等离子体 unmagnetized plasma
非定向的 omnidirectional
非对称效应 asymmetric effect
非恒定的 unsteady
非火山地热区 non-volcanic geothermal region
非均匀层 heterosphere
非均质性 heterogeneity
非矿异常 non-ore anomaly
非零等离子体温度 non-zero plasma temperature
非逆平行磁场 non-antiparallel magnetic fields
非偏移吸收 non-deviative absorption
非热辐射 non-thermal radiation
非线性波 non-linear wave
非线性扫描 non-linear sweep
非相干回波 non-coherent echo
非相干散射 incoherent scattering
非相干散射雷达 incoherent scattering radar
非寻常波 extraordinary wave
非炸药震源 non-explosive source
非质子耀斑 non-proton flare
菲列罗公式 Ferreros formula
沸泥塘 boiling mud pool
沸泉 boiling spring
费米加速 Fermi acceleration
分版原图 flaps
分瓣投影 interrupted projection
分辨率 resolution
分布法则〖地球化学的〗 principle of geochemical distribution
分布式的 distributed
分层 layer
分层设色表 graduation of tints
分层设色法 hypsometric layer
分割 segmentation
分光反照率 spectral albedo
分光光度学 spectrophotometry
分光通道 spectral channel
分类 breakdown
分力 component of force
分立能通道 separate energy channel
分量 component
分裂 splitting
分裂参数 splitting parameter
分区量值地图 choroplethic map
分区密度地图 dasymetric map
分区统计图表法 chorisogram method
分区统计图法 choroplethic method
分色 colour separation
分析地球化学 analytical geochemistry
分析地图 analytical map
分形 fractal
分支比 branching ratio
分子氮 molecular nitrogen
分子扩散 molecular diffusion
分子量 molecular weight
分子氢 molecular hydrogen
分子氢光谱 molecular hydrogen spectrum

分子氢化学 molecular hydrogen chemistry
分子吸收带 molecular absorption band
分子氧 molecular oxygen
分子氧发射 molecular oxygen emission
分子氧光谱 molecular oxygen spectrum
丰度值 abundance value
风切变 wind shear
风琴管振型 organ-pipe mode
风蚀 wind erosion
风矢量 wind vector
峰值 peak
峰值高度 peak altitude
峰值加速度 peak acceleration
峰值浓度 peak concentration
峰值速度 peak velocity
峰值位移 peak displacement
锋面活动 frontal activity
锋面位置 frontal position
伏拉索夫方程 Vlaslov equation
浮雕影像地图 picto-line map
浮动天顶仪 floating zenith telescope
浮力频率 buoyancy frequency
符号 symbol
符号化 symbolization
符合计数器 coincidence counter
幅度闪烁 amplitude scintillation
辐射传输 radiative transfer
辐射带 radiation belt
辐射附着 radiative attachment
辐射复合 radiative recombination
辐射功率 radiant power
辐射计 radiometer
辐射冷却 radiation cooling
辐射耦合 radiation coupling
辐射强度 radiant intensity
辐射强度表 pyranometer
辐射收支 radiation budget
辐射寿命 radiative lifetime
辐射通量 radiant flux
辐射图型 radiation pattern
辐射温度 radiation temperature
辐射效应 radiation effect
辐射压 radiative pressure
辐射仪 actinograph
辐射转移方程 equation of radiative transfer
辐射阻尼 radiation damping
福布什下降 Forbush decrease
俯测频高图 topside ionogram
俯冲带 underthrust belt
俯角 angle of depression
俯仰轴 pitch axis
腐殖质 humus
负相关 anticorrelation; inverse correlation; negative correlation
负异常 negative anomaly
附加位 additional potential
附体激波 attached shock wave
附着激波 attached shock wave
附着系数 attachment coefficient
复电阻率法 complex resistivity method
复合辐射 recombination radiation
复合率 recombination rate
复合系数 recombination coefficient
复离子 complex ion
复折射指数 complex refractive index
副赤道带 subequatorial belt
副极地带电离层 subpolar zone ionosphere
副极光带 subauroral zone
副极光带纬度 subauroral latitude
副极光区 subauroral region
副平流层 substratosphere
副热带 semi-tropical zone
傅里叶分析 Fourier analysis
覆被 cover
覆盖区 area coverage

改正 correction
钙地球化学障 calcic geochemical barrier
钙谱斑 calcium plage
盖层压力 overburden pressure
概率分布 probability distribution
概率误差 probable error
干旱区 arid area
干模式 dry model
干热岩体 hot dry rock
干涉滤波器 interference filter
干涉滤光片 interference filter
感生阻力 induction drag
感受效果 perceptual effect
感应测井 induction logging
感应脉冲瞬变法 INPUT method (induced pulse transient method)
感知分组 perceptual grouping
刚度谱 rigidity spectrum
高层大气 upper atmosphere
高层大气物理学 upper atmospheric physics
高层云 altostratus cloud
高程 elevation; height
高程测量 height measurement
高电离层风 high ionospheric wind
高度间隔 height interval
高度角 altitude angle; elevation angle
高分辨率 high resolution
高混杂频率 upper hybrid frequency
高阶近似 higher-order approximation
高阶振型 higher mode
高精度 high precision
高空大气学 aeronomy
高空风 high atmosphere wind
高空观测 upper air observation
高空核爆炸 high-altitude nuclear explosion
高空流线 upper level streamline
高空气候学 aeroclimatology
高空气象学 aerology
高空生物学 aerobiology
高空图解 aerogram
高能粒子 high energy particle
高频 high frequency
高气压系统 high pressure system
高斯波束 Gaussian beam
高斯分布 Gaussian distribution
高斯-克吕格投影 Gauss-Kruger projection
高斯平面子午线收敛角 Gauss grid convergence
高斯平面坐标 Gauss plane coordinate
高斯期 Gauss epoch
高斯投影方向改正 arc-to-chord correction in Gauss projection
高斯投影距离改正 distance correction in Gauss projection
高斯中纬度公式 Gauss mid-latitude formula
高速等离子体 high-speed plasma
高纬沉降区 high-latitude precipitation region
高纬黑子 high-latitude spot
高压脊 anticyclonic ridge
高曳力区 high drag region

高中层 upper mesosphere
格局 pattern
格局变化 pattern change
格拉芬堡台阵 Graefenberg array
格拉姆磁间段 Graham magnetic interval
格式 format
格网单元 cell
个别元素地球化学 geochemistry of individual elements
各向同性粒子 isotropic particle
各向同性通量 isotropic flux
各向异性传导率 anisotropic conductivity
各向异性的 anisotropic
各向异性电导率 anisotropic conductivity
各向异性电介质 anisotropic medium
各向异性散射 anisotropic scattering
根际 soil rhizosphere
根圈 soil rhizosphere
跟踪 tracking
跟踪数字化 tracing digitizing
耕作 tillage
工程测量 engineering survey
工程地震(学) engineering seismology
工作流 workflow
弓激波 bow shock
公元 Anno Domini
功函数 work function
功率密度谱 power density spectrum
功率谱 power spectrum
攻角 angle of attack
供电电极 current electrode
拱线运动 apsidal motion
拱心角 apsidal angle
共大地水准面 co-geoid
共轭光电子 conjugate photoelectron
共面轨道 coplanar orbit
共深度点叠加 CDP stacking (common-depth-point-stacking)
共深度点网格 CDP grid (common-depth-point grid)
共旋磁力线 corotating magnetic field line
共旋电场强度 corotation field strength
共振辐射 resonance radiation
共振散射 resonant scattering
共振射电爆发 sympathetic radio burst
共振探针 resonance probe
共振线 resonance line
共中心点叠加 common mid-point stacking
沟壑 gully
构造地球化学异常 structural geochemical anomaly
构造地震 tectonic earthquake
构造活动 tectonic activity
构造物理学 tectonophysics
构造应力 tectonic stress
古地磁 palaeomagnetism
古地磁场 palaeomagnetic field
古地磁赤道 palaeogeomagnetic equator
古地磁方向 palaeomagnetic direction
古地磁极 palaeomagnetic pole
古地磁强度 palaeogeomagnetic intensity
古地磁学 palaeomagnetism
古地热系统 ancient geothermal system; fossil geothermal system
古地热学 palaeogeothermics
古地图 ancient map
古典极光带 classical auroral zone
古经度 palaeolongitude
古水 fossil water
古纬度 palaeolatitude
固定辐射点 stationary radiant
固定台 base station
固体潮 earth tide
固体地球物理学 solid earth geophysics

固有振动 natural oscillation
拐角频率 corner frequency
观测 observation
观测集 observation set
观测误差 observational error
观测值 observed value
官方地形制图信息系统 Authoritative Topographic Cartographic Information System
贯穿辐射 penetrating radiation
贯穿深度 penetration depth
冠状日珥 cap prominence
惯性导航 inertial navigation
惯性飞行 coasting flight
惯性漂移 inertia drift
惯性轴 inertial axis
惯性坐标系 inertial coordinate system
灌溉 irrigation
光泵磁力仪 optical pump magnetometer
光程 optical distance
光电光度计 photoelectric photometer
光电离截面 photoionization cross section
光电流 photoelectric current
光电效应 photoelectric effect
光电子光谱学 photoelectron spectroscopy
光度曲线 luminosity curve
光化吸收 actinic absorption
光化效应 photochemical effect
光化学 photochemistry
光化学定律 laws of photochemistry
光化学反应 photochemical reaction
光化作用 photochemical action
光解作用 photolysis
光年 light year
光谱反照率 spectral albedo
光谱区 spectral region
光谱通道 spectral channel
光谱项 spectral term
光谱学 spectroscopy
光球模型 model photosphere
光通量密度 luminous flux density
光学窗 optical window
光学厚度 optical thickness
光压 light pressure
光栅 raster
光栅光谱 grating spectrum
光栅摄谱仪 grating spectrograph
光致电离 photoionization
光致复合 photo-recombination
光致激发 photo-excitation
光致离解 photodissociation
光致脱离 photodetachment
广角共深度点 WACDP (wide angle common depth point)
广角望远镜 wide-angle telescope
广角物镜 wide-angle object glass
广延相干簇射 extensive coherent shower
广义欧姆定律 generalized Ohm's law
广义射线 generalized ray
广义射线理论 GRT (generalized ray theory)
归航信标 homing beacon
归化纬度 reduced latitude
归心改正 correction for centring
归心元素 element of centring
归一化重力总梯度 normalized total gravity gradient
规划地图 planning map
规矩线 register mark
轨道 orbit
轨道角动量 orbital angular momentum
轨道衰变 orbital decay
轨道移动 orbital movement
轨道运动 orbital motion
轨道周期 orbital period
滚动轴 roll axis
滚轮式卫星 cartwheel satellite
国际参考大气 international reference atmosphere

国际参考地磁场 international geomagnetic reference field
国际参考电离层 international reference ionosphere
国际参考模式 international reference model
国际参考椭球 international reference ellipsoid
国际测绘联合会 International Union of Surveying and Mapping
国际磁层研究干事会 IMS Steering Committee
国际磁层研究公报 IMS Bulletin
国际磁层研究卫星形势中心 IMS Satellite Situation Centre
国际磁情记数 international magnetic character figure
国际单位制 International System of Units
国际地球物理年 International Geophysical Year
国际地震汇编 ISS (International Seismological Summary)
国际地震中心 ISC (International Seismological Center)
国际极光图 International Aurora Atlas
国际极年 International Polar Year
国际加速度计部署台网 IDA Network (International Deployment of Accelerometers Network)
国际宁静太阳年 International Quiet Sun Year
国际天文联合会 IAU (International Astronomical Union)
国际椭球体 international ellipsoid
国际原子时 IAT (International Atomic Time)
国际最扰日 international most disturbed days
国家地图集 national atlas
国家水准网 national leveling network
国土 land
过渡层 transition layer
过渡场法 transient field method
过渡带地震勘探 transitional zone seismic
过量电离 excess ionization
过热 overheating
过热光电子 super-thermal photoelectron
过水断面 flow cross-section
过氧化氢 hydrogen peroxide

Hh

哈拉米略事件 Jalamillo event
哈朗间断 Harang discontinuity
海岸带 coastal zone
海岸线 coastline
海槽 trough
海底波 water bottom event
海底地震仪 ocean-bottom seismograph; submarine seismograph
海底扩张 sea floor spreading
海福德椭球 Hayford ellipsoid
海上地震采集 marine seismic acquisition
海上地震数据采集 marine seismic acquisition
海图 chart
海王星 Neptune
海卫二 Nereid
海卫一 Triton
海下地震 submarine earthquake
海啸 tsunami
海啸地震 tsunami earthquake
海洋大地测量学 marine geodesy
海洋地球化学探矿 marine geochemical prospecting
海洋地震调查 marine seismic survey
海洋地震漂浮电缆 marine seismic streamer
海洋地震剖面仪 marine seismic profiler
海洋地震拖缆 streamer
海洋反射地震调查 marine reflection seismic survey
海洋广角反射地震调查 marine wide-angle reflection seismic survey
海洋折射地震调查 marine refraction seismic survey
海洋重力测量 gravity measurement at sea
海洋重力仪 sea gravimeter
海震 seaquake; sea shock
氦离子 helium ion
含氢物质 hydrogenous material
含沙量 sediment concentration
含水饱和度 water saturation
含水层 aquifer
含盐量 salt content
含油饱和度 oil saturation
航测 aerial survey
航道 channel
航迹恢复 flight-path recovery
航空磁测 aeromagnetic survey; airborne magnetic survey
航空磁测的 aeromagnetic
航空弹道学 aeroballistics
航空的 aerial; airborne
航空地球化学勘探 aerogeochemical prospecting; airborne geochemical prospecting
航空电磁法 AEM method (airborne electromagnetic method)
航空电磁系统 AEM system (airborne electromagnetic system)
航空放射性测量 airborne radioactivity survey
航空摄影机 aerial camera
航空生物学 aerobiology
航空天文学 aviation astronomy
航空图 aeronautical chart

航空遥感 aerial remote sensing
航空影像 aerial image
航空重力测量 aerial gravity measurement; airborne gravity measurement
航空重力仪 airborne gravimeter
航路指南 SD (sailing direction)
航摄计划 flight plan of aerial photography
航摄领航 navigation of aerial photography
航摄漏洞 aerial photographic gap
航摄软片 aerial film
航摄像片 aerial photograph
航摄质量 quality of aerophotography
航天飞机 space shuttle
航天摄影 space photography
航天摄影测量 space photogrammetry
航天学 astronautics
航天遥感 space remote sensing
航向 course
航向重叠 end overlap; fore-and-aft overlap; forward overlap; longitudinal overlap
航向倾角 longitudinal tilt; pitch
航行通告 notice to navigator
航行图 sailing chart
航行障碍物 navigation obstruction
耗尽层 depletion layer
耗散 dissipation
合并效率 merging efficiency
合成地图 synthetic map
合成地震图 synthetical seismogram
合成孔径雷达 SAR (Synthetic Aperture Radar)
合成孔径雷达图像 SAR image
合点控制 vanishing point control
合线 vanishing line
和达图 Wadati diagram
河道 river channel
河道整治测量 river improvement survey
河流域 river basin
河外宇宙线质子 extragalactic cosmic ray proton
河外致密射电源 extragalactic compact radio source
河网 river network
核磁共振 NMR (nuclear magnetic resonance)
核磁共振测探 NMR sounding
核地球化学 nuclear geochemistry
核点 epipole
核反应 nuclear reaction
核-幔边界 CMB (core-mantle boundary)
核-幔耦合 core-mantle coupling
核面 epipolar plane
核乳胶 nuclear emulsion
核天体物理学 nuclear astrophysics
核线 epipolar line; epipolar ray
核线相关 epipolar correlation
盒式分类法 box classification method
黑洞 black hole
黑体 black body
黑体光子积分通量 integrated blackbody photon flux
黑土 black soil
黑子磁场 spot magnetic field
黑子辐射 sunspot radiation
黑子活动性 sunspot activity
黑子极大值 spot maximum; sunspot maximum
黑子极小值 spot minimum; sunspot minimum
黑子极性 sunspot polarity
黑子日珥 sunspot prominence; prominence of the sunspot type
黑子双周 double sunspot cycle
黑子型日斑 prominence of the sunspot type
黑子耀斑 sunspot flare
黑子周 sunspot cycle

黑子周变化 sunspot cycle variation
痕量成分 trace constituent
恒定高度 constant altitude
恒时钟 sidereal clock
恒星际空间 interstellar space
恒星年 sidereal year
恒星日 sidereal day
恒星摄影机 stellar camera
恒星时 sidereal time
恒星视差 stellar parallax
恒星中天测时法 method of time determination by star transit
恒星钟 sidereal clock
横波 transverse wave
横波型面波 surface S wave
横断面测量 cross-section survey
横断面图 cross-section profile
横向测井 electrical lateral curve logging
横向传播 transverse propagation
横向的 lateral
横向电测井 electrical lateral curve logging
横向电导率 transverse conductivity
横向漂移 transverse drift
横轴投影 transverse projection
横坐标 abscissa
烘烤接触检验 baked contact test
红斑 red spot
红弧 red arc
红极光 red aurora
红外测距仪 infrared EDM instrument
红外大气光谱带系 infrared atmospheric band system
红外辐射计 infrared radiometer
红外辐射通量 infrared flux
红外亮温 infrared brightness temperature
红外片 infrared film
红外扫描仪 infrared scanner
红外摄影 infrared photography
红外天文学 infrared astronomy
红外图像 infrared imagery
红外遥感 infrared remote sensing
红移 red shift
宏观地震资料 macroseismic data
洪峰 flood peak
洪水 flood
洪水流量 flood discharge
后方交会 resection
后生地球化学异常 epigenetic geochemical anomaly
后随黑子 following sunspot
呼吸速率 respiration rate
弧度测量 arc measurement
弧光放电 arc discharge
湖泊测量 lake survey
湖震 seismic seiche
互补色地图 anaglyphic map
互补色镜 anaglyphoscope
互补色立体观察 anaglyphical stereoscopic viewing
互操作 interoperability
互换点 interlocking point
互扰日珥 interacting prominence
互相关 cross correlation
花样叠加 diversity stack
"花园门"悬挂法 garden gate suspension
滑动 slip
滑动函数 slip function
滑动角 rake
滑动接触法测井 scratcher electrode logging
滑动向量 slip vector
滑坡 landslide
滑塌地震相 slump seismic facies
滑翔飞行 coasting flight
化石磁化 fossil magnetization
化石磁化强度 fossil magnetization
化学层 chemosphere
化学层顶 chemopause
化学地球温度计 chemical geothermometer
化学发光 chemiluminance

化学丰度 chemical abundance
化学及动力学寿命 chemical and dynamic lifetime
化学平衡 chemical equilibrium
化学清洗 chemical cleaning
化学剩磁 CRM (chemical remanent magnetization)
还原地球化学障 reduction geochemical barrier
环电流 ring current
环电流强度 strength of the ring current
环境地球化学 environmental geochemistry
环境地图 environmental map
环境化学 environmental chemistry
环境生物地球化学 environmental biogeochemistry
环境探测卫星 environmental survey satellite
环境卫星 environmental satellite
环境有机地球化学 environmental organic geochemistry
环食 annular eclipse
环太平洋地震带 Circum-Pacific Seismic Zone
环天顶弧 circumzenithal arc
环形测深 loop-shaped sounding
环形场 toroidal field
环形低地 walled plain
环形对流 cellular convection
环型 toroidal
环型振荡 toroidal oscillation
缓冲区 buffer
缓和曲线测设 spiral curve location; transition curve location
缓始 emersio
缓始磁暴 gradual commencement (magnetic) storm
换能器 transducer
换能器吃水改正 correction of transducer draft
换能器动态吃水 transducer dynamic draft
换能器基线 transducer baseline
换能器基线改正 correction of transducer baseline
换能器静态吃水 transducer static draft
荒漠 desert
荒漠化 desertification
黄道带 zodiac
黄道光 zodiacal light
黄道面 ecliptic plane
黄道圈 zodiacal circle
黄道坐标 ecliptic coordinate
黄昏区段 evening sector
黄极 pole of the ecliptic
黄经 astronomical longitude; celestial longitude
黄土 loess
黄土高原 loess plateau
黄土丘陵 loess hill
黄土丘陵区 loess hilly region
黄纬 astronomical latitude; celestial latitude
灰色调 middle tone
灰体大气 gray atmosphere
灰楔 grey wedge
恢复相 recovery phase
回返 inversion
回光灯 signal lamp
回归方程 regression equation
回归分析 regression analysis
回归估计 regression estimation
回声测深 echo sounding
回声测深仪 echo sounder
回头曲线测设 hair-pin curve location
回旋共振 gyroresonance
回转波 reverse branch
汇聚带 convergence belt; convergence zone
汇聚型地热带 convergent-type geothermal belt
汇水面积测量 catchment area survey
会合太阴周 synodic lunar period

会合卫星 synodic satellite
会合运动 synodic motion
会合自转 synodic rotation
会聚效应 focusing effect
绘图机 plotter
绘图文件 plotting file
彗星群 group of comets
彗星族 family of comets
浑浊因子 turbidity factor
混波器 mixer
混合潮港 mixed tidal harbor
混合改正 complex correction
混合共振 hybrid resonance
活动断层 active fault
活动光带 activated band
活动黑子日珥 active sunspot prominence
活动极光 active aurora
活动日珥区 active prominence region
活动太阳现象 active sun phenomena
活化能 activation energy
活跃期 active phase
火成的 igneous
火成岩 igneous rock
火箭发射场 rocket-launching site
火箭气球装置 rocket balloon instrument
火箭推进剂 rocket propellant
火箭推进器 rocket propellant
火流星 bolide
火面学 areography
火球 bolide
火山 volcano
火山地热区 volcano-geothermal region
火山地震 volcanic earthquake
火山活动 volcanic activity
火山丘地震相 volcanic mound seismic facies
火山水 volcanic water
火山岩 volcanic rock
火山作用 volcanism
火卫二 Deimos
火卫一 Phobos
火星 Mars
火星表面学 areography
火星尘暴 Mars dust storm
火星电离层 Mars ionosphere
火星极盖 Mars polar cap
火星日气晖 Mars dayglow
火星逃逸速度 Mars escape velocity
火星物理学 areophysics
火星云 Mars cloud
火星震 Marsquake
火星中心的 areocentric
霍尔电导率 Hall conductivity
霍夫函数 Hough function

击穿 breakdown
机场测量 airport survey
机场跑道测量 airfield runway survey
机械剩磁 mechanical remanence
机械投影 mechanical projection
机载的 airborne
机载激光测深 airborne laser sounding
机载激光测探 airborne laser sounding
机载遥感器 airborne sensor
机助测图 computer-aided mapping; computer-assisted plotting
机助地图制图 CAC (computer-aided cartography)
机助分类 computer-assisted classification
机助制图 computer-aided mapping; computer-assisted plotting
积顶点〖地球化学的〗 geochemical culmination
积分电子浓度 integral electron concentration
积分密度谱 integral density spectrum
积分谱 integrated spectrum
积分通量 integral flux
积云线 cumulus cloud line
基本比例尺 basic scale
基本地球化学图 basic geochemical map
基本天文点 fundamental astronomical point
基本图形元素 primary graphic element
基本重力点 basic gravimetric point
基尔霍夫积分偏移 Kirchhoff integration migration
基-高比 base-height ratio
基频 fundamental frequency
基态 ground state
基线 baseline
基线测量 baseline measurement
基线飞行 baseline flying
基线网 baseline network
基亚曼间段 Kiaman interval
基岩 bedrock
基于GPS的 GPS-based
基准 datum
基准点 fiducial point
基准面静校正 datum static correction
基准台 standard station; track station
基准纬度 latitude of reference
畸变波 distorted wave
激波 shock wave
激波风洞 shock tunnel
激波谱 shock spectrum
激波强度 shock strength
激发极化测深 sounding of induced polarization
激发极化法 IP method (induced polarization method)
激发态 excited state
激光测高仪 laser altimeter
激光测距仪 laser ranger
激光测深仪 laser sounder
激光测月 LLR (lunar laser ranging)

激光地形仪 laser topographic position finder
激光二极管 LD (laser diode)
激光绘图机 laser plotter
激光经纬仪 laser theodolite
激光目镜 laser eyepiece
激光扫平仪 laser swinger
激光水准仪 laser level
激光投点 laser plumbing
激光照排机 image setter
激光准直法 method of laser alignment
激光准直仪 laser aligner
激励频率 driving frequency
吉尔伯特反极性期 Gilbert reversed polarity epoch
吉尔伯特期 Gilbert epoch
吉尔绍事件 Gilsa event
级联簇射 cascade shower
级联偏移 cascade migration
极大日冕 maximum corona
极地簇射 polar shower
极风 polar wind
极盖极光 polar cap aurora
极盖气辉 polar cap glow
极盖吸收 PCA (polar cap absorption)
极光 aurora
极光带 auroral belt
极光带电集流 auroral electrojet
极光带电集流指数 AE index (auroral electrojet index)
极光电子 auroral electron
极光发光 auroral luminescence
极光高度 height of aurora
极光光谱 auroral spectrum
极光弧 auroral arc
极光回波 auroral echo
极光活动 aurora activity
极光雷达 auroral radar
极光粒子 auroral particle
极光卵形环 auroral oval
极光冕 auroral corona
极光谱线 auroral line
极光千米波辐射 AKR (auroral kilometric radiation)

极光物理学 auroral physics
极光形式 auroral pattern
极光形态学 auroral morphology
极光云 auroral cloud
极光质子 aurora proton
极化效应 polarization effect
极距 codeclination
极浅海地震勘探 extreme shallow water seismic
极区磁扰 polar magnetic disturbance
极区等离子体 polar plasma
极区流量管 polar flux tube
极射线 polar ray
极隙 cleft; cusp
极限误差 limit error
极向水平 poleward horizon
极相漂移 polar phase shift
极型 poloidal
极型振荡 poloidal oscillation
极性超代 polarity superchron
极性过渡 polarity transition
极性间段 polarity interval
极性年代 polarity chron
极性年代测定 polarity dating
极性偏向 polarity bias
极性期 polarity epoch
极性事件 polarity event
极性序列 polarity sequence
极性亚代 polarity subchron
极移 polar motion; polar wander
极移路径 PWP (polar-wander path)
极移曲线 polar-wander curve
极震区 meizoseismal area
极轴 polar axis
极紫外线 extreme ultraviolet line
极坐标 polar coordinate
极坐标定位 point coordinate positioning
极坐标定位方法 azimuth distance positioning method; polar positioning method
极坐标缩放仪 polar pantograph

急流带　jet stream zone
急流强度　jet stream intensity
急脉冲　sudden impulse
急始　sudden commencement; sudden onset
急始磁暴　sudden commencement (magnetic) storm
集体效应　collective effect
集约效应　collective effect
几何大地测量学　geometric geodesy
几何定向　geometric orientation
几何反转原理　principle of geometric reverse
几何校正　geometric correction; geometric rectification
几何扩散　geometric spreading
几何模型　geometric model
几何条件　geometric condition
计曲线　index contour
计算机兼容磁带　CCT (computer compatible tape)
计算机视觉　computer vision
计算机制图综合　computer cartographic generalization
纪年法　chronology
季节效应　seasonal effect
既有线站场测量　survey of existing station yard
寂静地震　silent earthquake
加常数　addition constant
加利津地震仪　Galitzin seismograph
加密点　pass point
加权叠加　weighted stack
加速度反应谱　acceleration response spectrum
加速度计　accelerometer
加速度势　acceleration potential
加速度仪　accelerograph
加速机制　accelerating mechanism
加栅电离室　gridded chamber
夹层结构　sandwich construction
伽利略卫星　Galilean satellites
甲烷　methane
甲烷生成　methane production
假捕获粒子　pseudo-trapped particle
假彩色合成　false colour composite
假彩色片　colour infrared film
假彩色摄影　false colour photography
假彩色图像　false colour image
假定经度　assumed longitude
假定纬度　assumed latitude
假定坐标系　assumed coordinate system
假黄道光　false zodiacal light
假余震　pseudo-aftershock
尖点　cusp
尖端效应　effect of point
间曲线　half-interval contour
监测台　monitor station
监测网　monitoring network
监督分类　supervised classification
监视记录　monitor record
监视台　check station
检波器　geophone
检波器排列　spread
检测台　check station
检疫锚地　quarantine anchorage
减色印刷　reducing colour printing
减速辐射　deceleration radiation
减灾　disaster mitigation
剪辑　clipping
剪切波　shear wave
剪切不稳定性　shear instability
剪切耦合 PL 波　shear coupled PL waves
剪切熔融　shear melting
剪切位错　shear dislocation
剪切效应　shear effect
简谐波　simple harmonic wave
简正振型　normal mode
碱解　alkaline hydrolysis
碱金属　alkali metal
碱金属原子　alkali metal atom

碱性地球化学障 alkaline geochemical barrier
间隔速度 interval velocity
间接法纠正 indirect scheme of digital rectification
间接高空分析 indirect aerological analysis
间歇泉 geyser; intermittent spring
间歇泉区 geyserland
建设性干扰 constructive interference
建设用地 construction land
建造平均方向 formation mean direction
建筑工程测量 building engineering survey
建筑抗震 building aseismicity
建筑摄影测量 architectural photogrammetry
建筑物沉降观测 building axis survey
渐长区间 projection interval
渐长纬度 meridional part
渐近方向 asymptotic direction
渐近经度 asymptotic longitude
渐近纬度 asymptotic latitude
键联 binding
箭头图 arrow plot
箭载气象仪 rocket meteorograph
箭载质谱计 rocket-borne mass spectrometer
江河测量 river survey
江河图 river chart
降交点 southbound node
降频扫描 down sweep
降水 precipitation
降雨 rainfall
交变场清洗 AF cleaning (alternating field cleaning)
交叉耦合效应 cross-coupling effect
交叉调制 cross-modulation
交叉网线 cross-ruling
交点月 draconic month
交换不稳定性 interchange instability
交会 intersection
交流退磁 AC demagnetization (alternating current demagnetization)
交替极大值 alternate maxima
交替极小值 alternate minima
交线条件 Czapski condition; Scheimpflug condition
交向摄影 convergent photography
焦耳耗散 Joule dissipation
焦耳加热 Joule heating
焦距 focal length
焦面快门 curtain shutter
礁石 rock
角加速度 angular acceleration
角视立体图 corner cube display
校正 correction
校准 calibration
较软能谱 softer spectrum
较硬能谱 harder spectrum
教学地图 school map
接触激发极化法 contact induced polarization method
接触晒印 contact printing
接触印刷 contact printing
接纳锥 acceptance cone
接收 receiving
接收点静校正 receiver statics
接收二极管 reception diode
接收中心 receiving centre
节面 nodal plane
杰弗里斯-布伦走时表 Jeffreys-Bullen travel time table
结点平差 adjustment by method of junction point
结合 binding
结合能 binding energy
结合性分离 associative detachment
结晶剩磁 crystallization remanence; crystallization remanent magnetization
截断误差 truncation error
截面差改正 correction from normal section to geodetic

截止刚度 cut-off rigidity
解耦 decoupling
解析测图 analytical mapping
解析测图仪 analytical plotter
解析定向 analytical orientation
解析纠正 analytical rectification
解析空中三角测量 analytical aerotriangulation
解析摄影测量 analytical photogrammetry
解析图根点 analytic mapping control point
解阻场 unblocking field
解阻温度 unblocking temperature
介电测井 dielectric logging
介质 medium
介子生成层 meson producing layer
介子望远镜 meson telescope
界面波 boundary wave
界面速度 boundary velocity
界址点 boundary mark; boundary point
金属弹簧重力仪 metallic spring gravimeter
金星 Venus
金星环流 Venus circulation
金星灰光 ashen light
金星震学 Venus seismology
金字塔 pyramid
津格尔测时法 method of time determination by Zinger star-pair
进动 processional motion
近场 near field
近场地震学 near-field seismology
近程定位系统 short-range positioning system
近地点 perigee
近地点角距 argument of perigee
近地空间 terrestrial space
近红外 near-infrared
近极地轨道 near polar orbit
近井点 point for shaft position
近景摄影测量 close-range photogrammetry
近似地形面 telluroid
近似平差 approximate adjustment
近心点 pericenter
近星点 periastron
近月点 pericynthion
近震 near earthquake
近紫外 near ultraviolet
近紫外激发 near UV excitation
劲度 stiffness
禁航区 prohibited area
禁戒跃迁 forbidden transition
禁锚区 anchorage-prohibited area
禁区界限 forbidden zone boundary line
经度采用值 adopted longitude
经度差 difference of longitude
经度起算点 origin of longitude
经度效应 longitude effect
经济地图 economic map
经纬网格 fictitious graticule
经纬仪 theodolite
经纬仪测绘法 mapping method with transit
经纬仪导线 theodolite traverse
经向范围 longitudinal extent
经向环流 meridional circulation
晶体分光计 crystal spectrometer
精度 precision
精码 P Code (Precise Code)
精密测距 precise ranging
精密垂准 precise plumbing
精密导线测量 precise traversing
精密工程测量 precise engineering survey
精密工程控制网 precise engineering control network
精密估计 precision estimation
精密机械安装测量 precise mechanism installation survey
精密立体测图仪 precision stereoplotter
精密水准测量 precise leveling

精密水准仪 precise level
精密星历 precise ephemeris
精密准直 precise alignment
井底车场平面图 shaft bottom plan
井间地震 crosshole seismic
井径测井 caliper survey
井口时间 uphole time
井区 well zone
井上下对照图 surface-underground contrast plan
井深工程测量 shaft prospecting engineering survey
井田区域地形图 topographic map of mining area
井筒十字中线标定 setting-out of cross line through shaft centre
井下测量 underground survey
井下空硐测量 underground cavity survey
井眼 borehole
井液测井 well fluid logging
井中-地面方式 borehole-surface variant
井中电视 borehole televiewer
井中-井中方式 borehole-borehole variant
井中摄影 borehole photo
景观 landscape
景观地球化学 geochemistry of landscape
景观地球化学勘探 landscape geochemical prospecting
景观地图 landscape map
景观格局 landscape pattern
景观空间 landscape space
景观类型 landscape type
景观生态学 landscape ecology
径迹探测器 track detecter
径流 runoff
径矢 radius vector
径向分量 radial component
径向畸变 radial distortion
径向扩散 radial diffusion
径向矢量 radius vector
径向运动 radial motion

径向振荡 radial oscillation
净空区测量 clearance limit survey
静电复印 xerography
静电平衡 electrostatic equilibrium
静校正 static correction
静态定位 static positioning
静态机械放大倍数 statical mechanical magnification
静态遥感器 static sensor
静止参考系 rest frame
静止锋 stationary front
静止质量 rest mass
镜点场 mirror field
镜点电子 mirror electron
镜点高度 mirror altitude
镜点间距 mirror separation
镜点纬度 mirror latitude
镜反射 mirror reflection; specular reflection
镜像点 mirror point
纠正 rectification
纠正仪 rectifier
纠正元素 element of rectification
居里点 Curie point
居里温度 Curie temperature
局部大气 local atmosphere
局部异常 local anomaly
局部增温 localized heating
矩密度张量 moment-density tensor
矩形分幅 rectangular map subdivision
矩张量 moment tensor
矩震级 moment magnitude
距离方位定位 point coordinate positioning
距离-距离定位 range-range positioning
飓风云 hurricane cloud
聚集体 aggregate
聚焦效应 focusing effect
卷云状云 cirriform cloud
绝对定向 absolute orientation

绝对定向元素　element of absolute orientation
绝对航高　absolute flying height
绝对黑体　absolute black body
绝对校准　absolute calibration
绝对亮度　absolute brightness
绝对零度　absolute zero
绝对强度　absolute intensity
绝对闪烁效率　absolute scintillation efficiency
绝对摄动　absolute perturbation
绝对时间　absolute time
绝对太阳通量　absolute solar flux
绝对温度　absolute temperature
绝对涡度　absolute vorticity
绝对误差　absolute error
绝对星等　absolute magnitude
绝对阈　absolute threshold
绝对值　absolute value
绝对重力测量　absolute gravity measurement
绝对重力仪　absolute gravimeter
绝热加速　adiabatic acceleration
绝热调节　adiabatic adjustment
绝热再分布　adiabatic redistribution
军事工程测量　military engineering survey
军用地图　military map
军用海图　military chart
均方根误差　root-mean-square error
均衡异常　isostatic anomaly
均匀层　homosphere
均匀层顶　homopause
均匀磁化球　uniformly magnetized sphere
均匀等离子体　homogeneous plasma
均匀电场　uniform electric field
龟纹　moire
竣工测量　final construction survey

喀斯特 karst
卡埃纳事件 Kaena event
卡尔曼滤波器 Kalman filter
卡尼亚尔-德胡普法 Cagniard-De Hoop method [technique]
卡尼亚尔法 Cagniard method
开采沉陷观测 mining subsidence observation
开采沉陷图 map of mining subsidence
开窗 windowing
开窗检索 retrieval by window
开发地震 production seismic
开漂移轨道 open drift orbit
开位形 open configuration
勘测设计阶段测量 survey in reconnaissance and design stage
勘察 exploration
勘察地球化学 exploration geochemistry
勘探 exploration
勘探地球物理(学) exploration geophysics
勘探地震学 exploration seismology
勘探基线 prospecting baseline
勘探网测设 prospecting network layout
勘探线测量 prospecting line survey
勘探线剖面图 prospecting line profile map
康拉德界面 Conrad discontinuity; Conrad interface
康索尔海图 Consol chart
抗差估计 robust estimation
抗震的 earthquake-proof; shock resistant
抗震结构 earthquake-resistant structure
抗震墙 seismic structural wall
抗重力 antigravity
考古地磁(学) archaeomagnetism
考古摄影测量 archaeological photogrammetry
柯林电导率 Cowling conductivity
科奇蒂事件 Cochiti event
蝌蚪图 tadpole plot
壳分裂 shell splitting
壳下地震 subcrustal earthquake
可测太阳风 detectable solar wind
可见辐射 visible radiation
可见光谱 visible spectrum
可见极光 visible aurora
可控源地震学 controlled source seismology
可控震源 controlled source
可控震源法 vibroseis
可逆过程 reversible process
可视化 visualization
可视化技术 visualization technology
可视化研究 visualization research
可压缩流体 compressible fluid
克拉索夫斯基椭球 Krasovsky ellipsoid
克莱罗定理 Clairaut theorem
刻度 calibration
刻绘 scribing
刻图仪 scriber

坑道平面图 adit planimetric
坑探工程测量 adit prospecting engineering survey
空盒气压表 aneroid barometer
空盒气压计 aneroidograph
空化 cavitation
空基系统 space-based system
空间磁学 space magnetism
空间大地测量学 space geodesy
空间电荷波 space charge wave
空间电荷鞘 space charge sheath
空间电学 space electricity
空间分析 spatial analysis
空间格局 spatial pattern
空间光学 space optics
空间后方交会 space resection
空间化学 space chemistry
空间密度 space density
空间前方交会 space intersection
空间试验室 Spacelab
空间数据 spatial data
空间数据管理系统 spatial database management system
空间数据基础设施 SDI (spatial data infrastructure)
空间数据库 spatial database
空间数据转换 spatial data transfer
空间物理学 space physics
空间信息可视化 visualization of spatial information
空军大气模式 Air Force atmospheric model
空气波 air wave
空气动力的 aerodynamic
空气耦合瑞利波 air-coupled Rayleigh wave
空气枪 air gun
空气枪震源 air gun
空气污染 air pollution
空腔共振 cavity resonance
空区 gap
空中导线测量 aerophotogonometry
空中发射 air launch
空中三角测量 aerotriangulation

空中水准测量 aeroleveling
孔 aperture
孔径 aperture
孔径比 aperture ratio
孔径角 aperture angle
孔隙度 porosity
控制〖地球化学的〗 geochemical control
控制测量 control survey
控制点 control point
口径比 aperture ratio
枯落物 litter
枯水 low water
库仑定律 Coulomb's law
库仑碰撞 Coulomb collision
库仑阻尼 Coulomb damping
库容测量 reservoir storage survey
跨断层 cross-fault
跨河水准测量 river-crossing leveling
块体 block
块状图 block diagram
快粒子碰撞 fast particle impact
快门 shutter
快速等离子体流 stream of fast plasma
快速扩散 rapid diffusion
宽谱带 broad spectral band
宽线地震 wide line seismic
宽线剖面 wide line profile
矿产图 map of mineral deposits
矿场平面图 mining yard plan
矿床地球化学 geochemistry of mineral deposits
矿井测量 underground survey
矿区 mining area
矿区控制测量 control survey of mining areas
矿山 mine
矿山测量 mine survey
矿山测量交换图 exchanging document of mining survey
矿山测量图 mining map
矿山测量学 mine surveying
矿山经纬仪 mining theodolite

矿体几何学 mineral deposits geometry
矿体几何制图 geometrisation of ore body
矿异常 ore anomaly
框标 fiducial mark
框幅摄影机 frame camera
扩散平衡 diffusive equilibrium
扩散转印 diffusion transfer
扩展 F spread F
扩展地震剖面法 ESP (extended seismic profiling)
扩张极 pole of spreading
扩张率 spreading rate
扩张速率 spreading rate

拉科斯特悬挂法 LaCoste suspension
拉曼效应 Raman effect
拉莫尔频率 Larmor frequency
拉莫尔运动 Larmor motion
拉普拉斯点 Laplace point
拉普拉斯方位角 Laplace azimuth
拉尚漂移 Laschamp excursion
拉尚事件 Laschamp event
莱曼线 Lyman line
兰勃特等面积方位投影 Lambert projection
兰勃特投影 Lambert projection
蓝底图 blue key
蓝移 blue shift
朗道阻尼 Landau damping
勒夫波 Love wave
勒夫数 Love's number
雷暴电学 thunderstorm electricity
雷暴闪光 thunderstorm flash
雷暴事件 thunderstorm event
雷达测高仪 radar altimeter
雷达覆盖区 radar overlay
雷达极光 radar aurora
雷达信标 radar beacon
雷达应答器 radar responder
雷达指向标 radar ramark
雷诺数 Reynolds' number
累积持续时间 cumulative duration
累积辐射 integrated radiation
类木行星 Jovian planet
类星体 quasi-stellar object
类行星伴星 planetary companion
类型地图 typal map

冷等离子体 cold plasma
冷泉 cold spring
冷却过程 cooling process
冷焰 cold plume
冷阴极 cold cathode
离解 dissociation
离解电势 dissociation potential
离解电位 dissociation potential
离解性复合 dissociative recombination
离散波数法 discrete wavenumber method
离散波数有限元法 DWFE method (discrete wavenumber-finite element method)
离散极盖区极光 discrete polar cap aurora
离散谱 discrete spectrum
离心力 centrifugal force
离心力位 potential of centrifugal force
离心漂移 centrifugal drift
离源初动 anaseismic onset
离源角 take-off angle
离源震 anaseism
离子波 ion wave
离子对 ion pair
离子-分子动力学 ion-molecule kinetics
离子化学 ion chemistry
离子-离子中和 ion-ion neutralization
离子密度 ion density
离子浓度 ion concentration
离子迁移率 ion mobility
离子声不稳定性 ion-acoustic instability

离子团 ion cluster
离子-原子重新排列 ion-atom rearrangement
离子质谱计 ion mass spectrometer
黎明前增强 predawn enhancement
黎明子午线 dawn meridian
礼炮号航天站 Salyut Space Station
里氏震级 Richter magnitude
理论地球化学 theoretical geochemistry
理论地图学 theoretical cartography
理论地震图 theoretical seismogram
理论天体物理学 theoretical astrophysics
理论天文学 theoretical astronomy
理论最低低潮面 lowest normal low water
理论最高高潮面 highest normal high water
锂反应 lithium reaction
力场 field of force
力高 dynamic height
历史地球化学 historical geochemistry
历史地图 historic map
历史地震 historical earthquake
历史地震学 historical seismology
历史基线 historical baseline
历元 epoch
历元平极 mean pole of the epoch
立标 beacon
立井导入高程测量 induction height survey through shaft
立井定向测量 shaft orientation survey
立井激光指向（法） laser guide of vertical shaft
立体测图仪 stereoplotter
立体地图 relief map
立体观测 stereoscopic observation
立体观测模型 stereoscopic model
立体镜 stereoscope
立体判读仪 stereointerpretoscope
立体摄影测量 stereophotogrammetry
立体摄影机 stereocamera; stereometric camera
立体视觉 stereoscopic vision
立体像对 stereopair
立体坐标量测仪 stereocomparator
砾石检验 conglomerate test
粒子沉降带 particle precipitation zone
粒子辐射 particle radiation
粒子轰击 particle bombardment
粒子加速器测量 particle accelerator survey
粒子散射 particle scattering
粒子通量 particle flux
连结点 tie point
连续调 continuous tone
连续对比 successive contrast
连续发射 continuous emission
连续方式 continuous mode
连续减光板 continuous attenuator
连续谱 continuous spectrum
连续吸收 continuous absorption
连续注入 successive injection
帘幕式快门 focal plane shutter; curtain shutter
联测比对 comparison survey
联合平差 combined adjustment
联合剖面法 composite profiling method
联合震源定位 joint hypocentral determination
联络测线 crossline
联盟号宇宙飞船 Soyuz Spacecraft
联系测量 connection survey
联系三角形法 connection triangle method

联系数 correlate
链式反应 chain reaction
凉泉 cool spring
量测摄影机 metric camera
两侧地球化学障碍 bilateral geochemical barrier
两栖地震勘探作业 amphibious seismic operation
亮斑极光 patchy aurora
亮点 bright spot
亮度 lightness
亮度分布 brightness distribution
亮度系数 brightness coefficient
亮谱斑 bright flocculus
量底法 quantity base method
量子产额 quantum yield
钌 Ru (ruthenium)
烈度 intensity
烈度表 intensity scale
裂变材料 fissionable material
裂变产物 fission product
裂缝 fracture
裂缝观测 fissure observation
裂缝性的 fractured
裂隙 fracture
邻带方里网 grid of neighbouring zone
邻图拼接比对 comparison with adjacent chart
邻元法 neighbourhood method
林地 woodland
林地土壤 woodland soil
林分 forest stand
林业 forestry
林业测量 forest survey
林业基本图 forest basic map
临边昏暗 limb darkening
临界碰撞频率 critical collision frequency
临界频率 critical frequency
临震 impending earthquake
零长弹簧 zero-initial-length spring
零流速 zero flow velocity
零偏线 agonic line
零漂改正 correction of zero drift
零频地震学 zero-frequency seismology
零倾线 aclinic line
零时区 zero zone
零线改正 correction of zero line
零向量 B-axis; null vector
零相位效应 zero-phase effect
零重力 zero-g (zero gravity)
零子午线 zero meridian
领海基线测量 territorial sea baseline survey
留尼旺事件 Reunion event
流变性侵入体 rheological intrusion
流动等离子体 streaming plasma
流动状极光 streaming aurora
流截面 flow cross-section
流速 flow velocity
流速分布 flow velocity distribution
流体 fluid
流体静力学平衡 hydrostatic equilibrium
流体力学 fluid mechanics
流体性质 fluid property
流体运动轨迹 trajectory of fluid motion
流线分析 streamline analysis
流星爆发 meteor flare
流星尘埃 meteor dust
流星防护屏 meteor bumper
流星雷达 meteor radar
流星余迹 meteor trail
流域 basin
流域水文学 watershed hydrology
硫化氢气孔 putizze
硫质气孔 solfatara
六分仪 sextant
龙卷日珥 tornado prominence
隆起效应 bulge effect
鲁洛夫斯太阳棱镜 Roelofs solar prism
陆潮 earth tide
陆地 land

陆地卫星 Landsat
露点温度 dew point temperature
旅游地图 tourist map
滤波 filtering
滤光片 filter
卵形带 oval-shaped belt
掠入角 grazing incidence
掠射角 grazing angle
略最低低潮面 lower low water
轮胎形辐射带 doughnut-shaped zone
罗经标 compass adjustment beacon
罗兰-C定位系统 Loran-C positioning system
罗兰海图 Loran chart
罗盘经纬仪 compass theodolite
罗盘仪 compass
罗盘仪测量 compass survey
罗斯比波 Rossby wave
罗西-福勒表 Rossi-Forel scale
罗西-福勒烈度表 Rossi-Forel (intensity) scale
逻辑兼容 logical consistency
逻辑一致性 logical consistency
螺型位错 screw dislocation
螺旋轨道 spiral orbit
螺旋角 spiral angle
螺旋状结构 spiral structure
洛希极限 Roche limit
落后角 angle of lag
落角 angle of arrival
落球法 falling-sphere method

马赫数 Mach number
马默思事件 Mammoth event
麦德维捷夫-施蓬霍伊尔-卡尔尼克表 MSK (intensity) scale; Medvedev-Sponheuer-Karnik (intensity) scale
麦德维捷夫-施蓬霍伊尔-卡尔尼克烈度表 MSK (intensity) scale; Medvedev-Sponheuer-Karnik (intensity) scale
麦卡利表〖修订的〗 MM (intensity) scale; modified Mercalli (intensity) scale
麦卡利烈度表〖修订的〗 MM (intensity) scale; modified Mercalli (intensity) scale
麦克斯韦方程 Maxwell's equation
麦克斯韦分布 Maxwell distribution
脉冲重复频率 pulse recurrence frequency
脉冲反褶积 spike deconvolution
脉冲高度谱 pulse height spectrum
脉冲加热 impulse heating
脉冲频率 pulse frequency
脉冲整形 pulse shaping
脉动 fluctuating; pulsation; microseism
脉动暴 microseismic storm
脉动极光 pulsating aurora
脉动理论 pulsation theory
脉动学说 pulsation theory
脉高谱 pulse height spectrum
脉形甄别 pulse shape discrimination

慢度 slowness
慢度法 slowness method
慢漂移波 slow drift wave
芒塞尔色系 Munsell colour system
盲区 blind zone
锚地 anchorage
锚位 anchorage berth
卯酉面 prime vertical plane
卯酉圈 prime vertical
卯酉圈曲率半径 radius of curvature in prime vertical
冒汽地面 fumarolic field; steaming ground
没影线 vanishing line
煤田地震勘探 coal seismic prospecting
美国空军大气模式 Air Force atmospheric model
镁离子反应 Mg^+ reaction
蒙戈湖漂移 Mungo Lake excursion
蒙绘 mask artwork
蒙片 mask
米尔恩-萧地震仪 Milne-Shaw seismograph
密度测井 density logging
密度剖面 density profile
密切轨道 osculating orbit
幂次律蠕变 power-law creep
冕风 coronal wind
冕流 coronal streamer
面波 surface wave
面波震级 surface wave magnitude
面积定律 law of areas
面水准测量 area leveling

面状符号　area symbol
瞄直法　sighting line method
民用历　civil calendar
名义量表　nominal scaling
明礁　bare rock
鸣震　ringing
冥王星　Pluto
冥卫一　Charon
模糊度　ambiguity
模糊度解算　ambiguity resolution
模糊分类法　fuzzy classifier method
模糊影像　fuzzy image
模拟磁带　analog tape
模拟磁强计　analog magnetometer
模拟地图　analog map
模拟法测图　analog photogrammetric plotting
模拟空中三角测量　analog aerotriangulation
模拟立体测图仪　analog stereoplotter
模拟摄影测量　analog photogrammetry
模式　pattern
模式识别　pattern recognition
模型连接　bridging of model
模型缩放　scaling of model
模型置平　leveling of model
莫霍界面　M discontinuity (Mohorovicic discontinuity); Moho
莫霍洛维契奇界面　M discontinuity (Mohorovicic discontinuity); Moho
莫里陨石　Murray meteorite
莫洛坚斯基公式　Molodensky formula
莫洛坚斯基理论　Molodensky theory
莫诺湖漂移　Mono Lake excursion
墨卡托海图　Mercator chart
墨卡托投影　Mercator projection
默坎顿间段　Mercanton interval
母彗星　parent comet
木卫二　Europa
木卫六　Himalia
木卫七　Elara
木卫三　Ganymede
木卫四　Callisto
木卫五　Amalthea
木卫一　Io
木星　Jupiter
木星表面的　zenographic
木星红斑　Jupiter red spot
木星热辐射　Jupiter heat radiation
木星云带　Jupiter cloud belt
目标反射器　target reflector
目标区　target area
目视观测者　visual observer
目视极光　visual aurora
目视流星　visual meteor
目视判读　visual interpretation
目视天顶仪　visual zenith telescope

钠 Na (sodium)
钠层 sodium layer
钠原子发射 sodium atomic emission
钠云 sodium cloud
南北不对称 north-south asymmetry
南极臭氧分布 Antarctic ozone distribution
南极盖 southern polar cap
南极光 aurora australis
南偶极子极 south dipole pole
南向磁场 southward field
南向值 southward value
内禀磁要素 intrinsic magnetic element
内部定向 interior orientation
内辐射带 inner radiation belt
内核 inner-core
内极光带 inner auroral zone
内生蒸汽 endogenous steam
内行星 inferior planet; inner planet
内源场 internal field
内转换 internal conversion
能见度 visibility
能见敏锐度 visibility acuity
能量转化效率 energy conversion efficiency
能谱 energy spectrum
能谱测井 spectral logging
能谱硬性 hardness of the spectrum
泥浆 mud
泥浆测井 mud logging
泥泉 mud spring
泥沙 sediment
泥沙运动 sediment movement
拟断面图 pseudosection map
拟合 fitting
拟稳平差 quasi-stable adjustment
逆时针的 anticlockwise
逆温层 inversion layer
逆转点法 reversal points method
年代学 chronology
年均值 annual mean
年历 almanac
年轮气候学 tree-ring climatology
年平均海面 annual mean sea level
黏度耦合 viscous coupling
黏滑 stick slip
黏滞力 viscous force
黏滞剩磁 viscous remanence; VRM (viscous remanent magnetization)
黏滞性相互作用 viscous-like interaction
鸟瞰图 bird's eye view map
宁静初级辐射 quiescent primary radiation
宁静光谱 quiescence spectrum
宁静日珥 quiescent prominence
宁静太阳风 quiet solar wind
宁静条件 quiet condition
牛顿定律 Newton's law
扭曲不稳定性 kink instability
扭转角 angle of torsion
扭转通量管 twisted flux tube
扭转型 torsional
扭转型振荡 torsional oscillation
农田 farmland

农田土壤　farmland soil
农业地球化学　agricultural geochemistry

努尼瓦克事件　Nunivak event
暖等离子体　warm plasma
暖锋　warm front

欧弗豪泽磁力仪 Overhauser magnetometer
欧洲遥感卫星 ERS (Europe Remote Sensing Satellite)
偶发地磁暴 sporadic geomagnetic storm
偶极 dipole
偶极测深 dipole electrode sounding
偶极排列 dipole-dipole array; dipole electrode array
偶极排列法 dipole-dipole array method
偶极子 dipole
偶极子午线 dipole meridian
偶极子坐标 dipole coordinate
偶然计数 accidental count
偶然误差 accidental error
偶现流星 sporadic meteor
偶现日珥 incidental prominence
耦合离子质谱计 coupled ion mass spectrometer

帕克模式 Parker model
帕特森反向 Paterson reversal
拍频 beat frequency
排列 spread
排列系数 array factor
判读 interpretation
判读仪 interpretoscope
旁向重叠 lateral overlap; side overlap
旁向倾角 lateral tilt
旁折光差 horizontal refraction error
炮点静校正 shoot statics
炮检距 shot-geophone distance
配准 registration
喷气孔 fumarole
喷泉效应 fountain effect
喷射日珥 spray prominence
盆地 basin
膨胀 dilatancy
膨胀波 dilatational wave
膨胀-扩散模式 DD model (dilatancy-diffusion model)
膨胀相 expansive phase
膨胀仪 dilatometer
膨胀硬化 dilatancy hardening
碰撞电离 ionization by collision
碰撞强度 collision strength
碰撞阻尼 collisional damping
偏航 yaw
偏航角 yaw angle
偏航轴 yawing axis
偏角 angle of declination
偏角法 method of deflection angle
偏食 partial eclipse
偏向角 angle of deviation
偏心偶极子 eccentric dipole

偏移 migration; offset
偏移成像 migration imaging
偏移距 offset
偏移速度 migration velocity
偏移速度分析 migration velocity analysis
偏移吸收 deviative absorption
偏振辐射 polarization radiation
偏振光立体观察 vectograph method of stereoscopic viewing
偏振角 polarization angle
偏振面 plane of polarization
偏振椭圆 polarization ellipse
漂移 drift
漂移轨道 drift orbit
频标 frequency scale
频带 frequency band
频带展宽 bandspread
频段 frequency band
频率 frequency
频率波数偏移 F-W migration (frequency-wavenumber migration)
频率测深法 frequency sounding method
频率分布 frequency distribution
频率分配 frequency assignment
频率急偏 SFD (sudden frequency deviation)
频率漂移 frequency drift
频率误差 frequency error
频率域 frequency domain
频偏 frequency offset
频漂 frequency drift
频谱 frequency spectrum
频谱激发极化法 spectral induced polarization method

频散波 dispersion wave
频时图 frequency-time record; sonogram
频时图式 frequency-time pattern
平板仪 plane-table
平板仪测量 plane-table survey
平板仪导线 plane-table traverse
平层近似 flat-layer approximation
平差 adjustment
平差法 adjustment method
平差计算 adjustment calculation
平差模型 adjustment model
平差值 adjusted value
平点 flat spot
平衡潮 equilibrium tide
平衡能谱 equilibrium energy spectrum
平极 mean pole
平均地球椭球 mean earth ellipsoid
平均海面 mean sea level
平均海面归算 seasonal correction of mean sea level
平均碰撞时间 mean collision time
平均曲率半径 mean radius of curvature
平均太阳风速 average solar wind speed
平均误差 average error
平均响应 average response
平均运动 mean motion
平均滞后 average lag
平均自由程 mean free path
平流层 stratosphere
平流层变化 stratospheric variation
平流层顶 stratopause
平流层动力学 stratospheric dynamics
平流层风场 stratospheric wind field
平流层环流 stratospheric circulation
平流层结构 stratospheric structure
平流层逆温层 stratospheric inversion layer
平流层耦合 stratosphere coupling
平流层热场 stratospheric thermal field
平流层突然增温 sudden stratospheric warming
平面剪切裂纹 in-plane shear crack
平面控制点 horizontal control point
平面控制网 horizontal control network
平面偏振 plane polarization
平面曲线测设 plane curve location
平面图 plane
平面坐标 horizontal coordinate
平时钟 mean-time clock
平太阳日 mean solar day
平行电导率 parallel conductivity
平行圈 parallel circle
平移参数 translation parameter
平整土地测量 survey for land smoothing
屏蔽高度 screening height
屏蔽效应 shielding effect
屏幕地图 screen map
坡度 slope gradient
坡度测设 grade location
坡面 slope surface
坡面经纬仪 slope theodolite
坡向 slope aspect
破坏准则 failure criterion
破裂 rupture
破裂长度 rupture length
破裂传播 rupture propagation
破裂过程 rupture process
破裂前沿 rupture front
破裂准则 fracture criterion
剖面 profile
剖面图 profile map
普拉烈系统 PRARE (Precise Range and Rangerate Equipment)

普拉特-海福德均衡 Pratt-Hayford isostasy
普朗克定律 Planck law
普雷斯-尤因地震仪 Press-Ewing seismograph
普通地图 general map
普通地图集 general atlas
普通海图 general chart

谱斑状耀斑 plage flare
谱函数 spectral function
谱线分裂 line splitting
谱线强度 spectral line intensity
谱线位移 line shift
谱线系 spectral series
谱指数 spectral index

起始相 starting phase
气爆震源 gas exploder
气候的 climatological
气候学的 climatological
气辉 airglow
气辉发射 airglow emission
气辉亮度 airglow brightness
气球观测 balloon observation
气球卫星 balloon satellite
气体倍增因子 gas multiplication factor
气体地球化学测量 geochemical gas survey
气体电离室 gas ionization chamber
气体激光器 gas laser
气相滴定 gas-phase titration
气象代表误差 meteorological representation error
气象图 meteorological chart
气象学 meteorology
气旋发生 cyclogenesis
气旋生成 cyclogenesis
气压层 barosphere
气压层顶 baropause
气压计 barograph; barometer
气压记录器 barograph
气压图 barogram
汽孔 steam vent
恰可察觉差 JND (just noticeable difference)
恰普斯基条件 Czapski condition; Scheimpflug condition
千米尺 kilometer scale
迁移波状地震相 migrating wave seismic facies
迁移率 mobility coefficient
迁移气候系统 migrating weather system
迁移系数 mobility coefficient
铅垂线 plumb line
铅垂仪 plumb aligner
前导黑子 leading sunspot; preceding spot
前进波 progressive wave
前向重叠 forward overlap
前向散射 forward scatter
前兆 precursor
前兆的 precursory
前兆时间 precursor time
前震 foreshock
钱德勒摆动 Chandler wobble
钱德勒章动 Chandler wobble
潜波 diving wave
浅地层剖面仪 sub-bottom profiler
浅海地震勘探 shallow water seismic
浅海海底电缆 bay cable
浅滩 shoal
浅源地震 shallow-focus earthquake
浅震 shallow-focus earthquake
强爆发 intense burst
强地磁活动 strong geomagnetic activity
强地动 strong (ground) motion
强地动地震学 strong-motion seismology
强地面运动 strong (ground) motion
强锋带 strong frontal zone
强耦合 strong coupling

强迫函数　forcing function
强震　strong earthquake
强震仪　strong-motion seismograph
桥墩定位　location of pier
桥梁测量　bridge survey
桥梁控制测量　bridge construction control survey
桥梁轴线测设　bridge axis location
桥位地形图　topographic map of bridge site
切线支距法　tangent off-set method
切向畸变　tangential distortion; tangential lens distortion
侵入　invasion
侵蚀　erosion
轻离子　light ions
轻元素　light element
氢过氧自由基反应　hydroperoxyl radical reaction
氢谱斑　hydrogen flocculus
倾角　angle of inclination; dip; dip angle; inclination
倾角测井　dipmeter survey
倾角赤道　dip equator
倾向定向　dip orientation
倾斜　tilt
倾斜叠加　slant stack
倾斜度变化　obliquity change
倾斜改正　tilt correction
倾斜观测　tilt observation
倾斜时差校正　DMO (dip move-out)
倾斜位移　tilt displacement
倾斜仪　clinometer; tiltmeter
清绘　fair drawing
琼斯不稳定判据　Jeans instability criterion
琼斯长度　Jeans length
丘陵　hill
秋分（点）　autumnal equinox
求积仪　planimeter
球面发散补偿　spherical divergence compensation
球面角　spherical angle
球面投影　stereographic projection
球面坐标　spherical coordinate
球谐分析　spherical harmonic analysis
球谐项　spherical harmonic term
球谐展开　spherical harmonic expansion
球心投影　gnomonic projection
球型　spheroidal
球型振荡　spheroidal oscillation
球载探测器　balloon-borne detector
球载探针　balloon-borne sonde
球状闪电　ball lightning
区划地图　regionalization map
区域K指数　local K index
区域地球化学　regional geochemistry
区域地球化学背景　regional geochemical background
区域地球化学分异　regional geochemical differentiation
区域地球化学勘探　regional geochemical prospecting
区域地球化学异常　regional geochemical anomaly
区域地图集　regional atlas
区域地震　regional earthquake
区域地震地层学　regional seismic stratigraphy
区域地质调查　regional geological survey
区域地质图　regional geological map
区域网平差　block adjustment
区域异常　regional anomaly
曲率半径　radius of curvature
曲率漂移　curvature drift
曲率矢量　curvature vector
曲线光滑　line smoothing
圈闭　trap
圈状日珥　loop prominence
全波理论　full wave theory
全氮　total nitrogen

全反射 total reflection
全方位检波器组合 omnidirectional geophone pattern
全辐射 total radiation
全惯性制导 all-inertial guidance
全环食 total-annular eclipse
全景畸变 panoramic distortion
全景摄影 panoramic photography
全景摄影机 panorama camera
全能法测图 universal method of photogrammetric mapping
全球大气环流 global atmospheric circulation
全球导航卫星系统 GLONASS (Global Navigation Satellite System)
全球电位梯度 worldwide potential gradient
全球定位系统 GPS (Global Position System)
全球风系 global wind system
全球环流 global circulation
全球极光 worldwide aurora
全球气象学 global meteorology
全球热量收支 global heat budget
全球热平衡 global heat balance
全球数字地震台网 GDSN (Global Digital Seismograph Network)
全球图 global picture
全球性地热带 planet-wide geothermal belt
全色红外片 panchromatic infrared film
全色片 panchromatic film
全食 total eclipse
全食带 belt of totality; path of total eclipse; zone of totality
全食时间 duration of totality
全食终 end of totality
全天空光度计 all-sky photometer
全天空照相机 all-sky camera
全息摄影 hologram photography
全向的 omnidirectional
全新世 holocene
群速度 group velocity

燃尽高度 burnout altitude
扰动 disturbing
扰动场地方时不均匀性 disturbance local-time inequality
扰动矢量 disturbance vector
扰动位 disturbing potential
扰动系数 perturbation coefficient
扰动质量 disturbing mass
扰日日变化 disturbed daily variation
绕射 diffraction
热层 thermosphere
热层顶 thermopause
热层事件 thermospheric event
热传导方程 heat-conduction equation
热磁分离 thermomagnetic separation
热磁曲线 thermomagnetic curve
热带的 tropical
热带对流层 tropical troposphere
热带高对流层 upper tropical troposphere
热带红弧 tropical red arc
热带气辉 tropical airglow
热带气旋 tropical cyclone
热等离子体 thermal plasma
热等离子体离子 hot plasma ion
热等离子体漂移 thermal plasma drift
热点 hot spot
热电子气体 hot electron gas
热辐射 thermal radiation
热各向异性 thermal anisotropy
热壑 heat sink

热寂 heat death
热扩散 thermal diffusion
热力学第二定律 second law of thermodynamics
热力学过程 thermodynamic process
热力学函数 thermodynamical function
热量收支 heat budget
热流 heat flow
热流单位 HFU (heat flow unit)
热流区 heat flow province
热流亚区 heat flow subprovince
热能分布 thermal energy distribution
热能密度 thermal energy density
热膨胀 thermal expansion
热平衡 thermal equilibrium
热清洗 thermal cleaning
热剩磁 thermoremanence; TRM (thermoremanent magnetization)
热逃逸 thermal escape
热效应 thermal effect
热焰 hot plume
热源 heat source
热源实验 heating source experiment
热源特性 heat-source characteristics
热致潮汐 thermally-driven tide
热致潮汐运动 thermal tide motion
热致电离 thermal ionization
人工尘埃 man-made dust
人工磁化法 artificial magnetization method

人工地震 artificial earthquake
人工发射 artificial emission
人工林 plantation
人工震源 artificial seismic source
人机交互处理 interactive processing
人口地图 population map
人体生物伦琴当量 Roentgen-equivalent-man
人文地图 human map
人造彗星 artificial comet
人造极光 artificial aurora
人造气辉 artificial airglow
人造卫星 artificial satellite
刃型位错 edge dislocation
认知制图 cognitive mapping
任意投影 arbitrary projection
日本气象厅表 JMA (intensity) scale; Japan Meteorological Agency (intensity) scale
日本气象厅烈度表 JMA (intensity) scale; Japan Meteorological Agency (intensity) scale
日磁情指数 daily magnetic character figure
日地暴 solar-terrestrial storm
日地的 solar-terrestrial
日地空间 solar-terrestrial space
日地物理学 solar-terrestrial physics
日地线 sun-earth line
日珥 solar prominence
日珥分光镜 prominence spectroscope
日光强度自动记录器 actinograph
日晷投影 gnomonic projection
日环食 annular eclipse
日冕 solar corona
日冕加热机制 mechanism of coronal heating
日冕扰动 coronal disturbance
日面 solar disk
日面经度 solar longitude
日面图 heliographic chart
日面纬度 solar latitude
日面坐标 heliographic coordinate
日偏食 partial solar eclipse
日气辉 day airglow
日球层 heliosphere
日球层电流片 heliospheric current sheet
日全食 total solar eclipse
日食 eclipse of the sun
日食效应 eclipse effect; solar eclipse effect
日下点 sub-solar point
日下点密度隆起 sub-solar density bulge
日心距离 heliocentric distance
日心坐标 heliocentric coordinate
日月岁差 lunisolar precession
日照钡云 sunlit barium cloud
日照极光 sunlit aurora
容许谱线 permitted line
容许跃迁 permitted transition
铷蒸气磁强计 rubidium magnetometer
蠕变 creep
蠕变仪 creepmeter
蠕滑 creep
入射流 incident stream
入渗 infiltration
软X线通量 soft X-ray flux
软成分 soft-component
软电子总通量 total soft electron flux
软辐射 soft radiation
软粒子 soft particle
软流层 asthenosphere
锐边界 sharp boundary
瑞利波 Rayleigh wave
瑞利极限 Rayleigh limit

塞曼效应 Zeeman effect
三重符合 threefold coincidence
三极排列 pole-dipole array
三角测量 triangulation
三角点 triangulation point
三角高程测量 trigonometric leveling
三角高程网 trigonometric leveling network
三角锁 triangulation chain
三角网 triangulation network
三角学的 trigonometric
三角洲 delta
三体反应 three-body reaction
三体附着 three-body attachment
三体复合 three-body recombination
三维地景仿真 3-D terrain simulation; three-dimensional terrain simulation
三维地形 3-D terrain; three-dimensional terrain
三维地震法 3-D seismic method; three-dimensional seismic method
三维可视化 3-D visualization; three-dimensional visualization
三维偏移 3-D migration; three-dimensional migration
三维数据体 3-D data volume; three-dimensional data volume
三维显示 3-D display; three-dimensional display
三轴磁强计 triaxial magnetometer
散见E层 sporadic E
散射 scattering
散射γ测井 scattered γ-ray logging
散射几率 scattering probability
散射角 scattering angle
散射系数 scattering coefficient
扫描分光计 scanning spectrometer
扫描数字化 scan-digitizing
扫描微波谱仪 scanning microwave spectrometer
扫频测深仪 sweep frequency sounder
色彩管理系统 colour management system
色调 tone
色环 colour wheel
色球 chromosphere
色球高层 upper chromosphere
色球模型 model chromosphere
色球针状物 chromospheric spicule
色散关系 dispersion relation
色相 hue
森林 forest
沙漠的 sandy
沙漠 desert
沙漠化 desertification
沙丘 dune
砂 sand
砂岩 sandstone
砂岩储层 sandstone reservoir
晒版 plate copying; printing down
山地 mountain
山根 root of mountain
闪变弧 flickering arc
闪电活动 lightning activity
闪光谱 flash spectrum

闪烁体 scintillator
扇形边界 sector boundary
扇形结构 sector structure
熵波 entropy wave
上包络 upper envelope
上地幔 upper mantle
上古代大气 primitive atmosphere
上流激波 upstream shock
上流离子 upstreaming ion
上流离子流 upward ion current
上气层 metasphere
上升电子 upgoing electron
烧蚀 ablation
哨声 whistler
哨声波散 whistler dispersion
舍入误差 round-off error
设计谱 design spectrum
射程能量曲线 range-energy curve
射出红外辐射 outgoing infrared radiation
射电窗 radio window
射电等强线 radio-isophote
射电辐射 radio emission
射电极光 radio aurora
射电亮度 radio brightness
射电谱 radio spectrum
射电谱斑 radio plage
射电食 radioeclipse
射电通量 radio flux
射电星 radio star
射气测量 emanation survey
射束效率 beam efficiency
射线 ray
射线参数 ray parameter
射线法 ray method
射线方程 ray equation
射线极光弧 rayed arc
射线状结构 ray structure
射线追踪 ray tracing
摄影测量 photogrammetry
伸长仪 extensometer
伸展磁场 stretched-out field
深成水 plutonic water
深地震测深 deep seismic sounding
深度 depth
深度感 depth perception
深度偏移 depth migration
深度剖面 depth (record) section
深海地震勘探 deep water seismic
深空 deep space
深空探测器 deep space probe
深色调 shade
深源地震 deep-focus earthquake
深震 deep-focus earthquake
甚长基线干涉测量 VLBI (very long baseline interferometry)
甚低频 very low frequency
甚低频带辐射场系统 very low frequency band radiated field system
甚低频发射 VLF emission
甚低频法 VLF method (very low frequency method)
甚低频噪声 VLF noise
甚高频 very high frequency
渗流 seepage
升轨 ascending pass
升交点 ascending node
升频扫描 up sweep
生命线 lifeline
生热单位 heat generation unit
生物舱 biopak
生物地球化学 biogeochemistry
生物地球化学勘探 biogeochemical prospecting
生物地球化学性疾病 biogeochemical disease
生物地球化学异常 biogeochemical anomaly
生物地球化学障 biogeochemical barrier
生物航天学 bioastronautics
生物圈 biosphere
生物容器 biopak
生物宇宙航行学 bioastronautics
声波 acoustic wave
声波测井 acoustic logging; sonic logging
声频 acoustic frequency

声速　acoustic velocity
声学定位测距声道　SOFAR channel (sound fixing and ranging channel)
声重力波　acoustic-gravity wave
盛行风　dominant wind
剩磁年龄　age of remanence
剩磁通量密度　residual flux density
剩余磁化　remanence; remanent magnetization
剩余磁化强度　remanent magnetization
剩余射程　residual range
剩余重力异常　residual gravity anomaly
失重　zero-g (zero gravity)
失重的　agravic
施赖伯全组合测角法　Schreiber method in all combinations
施伦伯格电极排列　Schlumberger electrode array
施伦伯格排列　Schlumberger array
施密特照相机　Schmidt camera
湿地　wetland
湿模式　wet model
石漠　hammada; stony desert
石漠化　rocky desertification
石油勘探　oil exploration
石陨星　asiderite
时变比例　time variant scaling
时变滤波　time-variable filtering
时标　timing marker
时差　moveout; stepout
时号　timing marker
时间导数　time derivative
时间叠加分析　superposed-epoch analysis
时间分辨率　time resolution
时间脉冲　time pulse
时间平滑　time smoothing
时间剖面　time (record) section
时间切片　time slice
时间项　time-term
时间项法　time-term method
时距曲面　surface hodograph
时距曲线　T-X curve (time-distance curve)
时深转换　time depth conversion
时序规则性　temporal regularity
识别码　identification code
实测数据　measured data
实际场强　actual field intensity
实际地磁场　real geomagnetic field
实际电流体系　real current system
实时数据　real time data
实时相关　real time correlation
实时遥测　real time telemetry
实验地球化学　experimental geochemistry
实验室等离子体　laboratory plasma
实验线路板　breadboard
实用地图学　applied cartography
食带　path of eclipse; zone of eclipse
食相　phase of an eclipse
食效应　eclipse effect
示坡线　slope line
矢量　vector
矢量化　vectorization
矢量绘图　vector plotting
矢量数据　vector data
矢量图形　vector image
世界地图集　world atlas
世界范围标准地震台网　WWSSN (World Wide Standard Seismograph Network)
世界气候　world climate
势能　potential energy
视差　parallax
视场对比　simultaneous contrast
视磁化率　apparent magnetic susceptibility
视电阻率　apparent resistivity
视高　apparent height
视极移　apparent polar wander
视极移路径　APWP (apparent polar-wander path)

视极移曲线 apparent polar-wander curve
视觉变量 visual variable
视觉层次 visual hierarchy
视觉对比 visual contrast
视觉分辨敏锐度 resolution acuity
视觉立体地图 stereoscopic map
视觉平衡 visual balance
视频 video frequency
视速度 apparent velocity
视星等 apparent magnitude
视应力 apparent stress
视准仪 alidade
适宜性 suitability
适应性水平 adaptation level
释能过程 exoenergic process
守恒定律 conservation law
守恒量 conserved quantity
首波 head wave
首曲线 intermediate contour
受迫振荡 forced oscillation
受压磁层 compressed magnetosphere
舒曼共振 Schumann resonance
舒曼紫外 Schumann UV
输移 transport
输运方程 transport equation
属群流星 shower meteor
属性 attribute
属性检定 attribute testing
属性精度 attribute accuracy
曙暮光 twilight
曙暮光发射 twilight emission
曙暮光弧 twilight arc
曙暮光谱 twilight spectrum
曙暮气辉 twilight airglow
曙暮霞 twilight colour
束导 beam riding
束缚电荷 bound charge
束缚-自由跃迁 bound-free transition
数据标准 data standard
数据采集 data aquisition; data capture
数据仓库 data warehouse
数据处理 data processing
数据格式 data format; digital format
数据更新 data updating
数据简化 data reduction
数据解释 data interpretation
数据可视化 data visualization
数据库设计 database design
数据库系统 database system
数据压缩 data compression; data reduction
数据样品 data sample
数据质量控制 data quality control
数量感 quantitative perception
数学地图学 mathematical cartography
数字城市 digital city
数字地图 digital map
数字地图产品标准 product standard of digital maps
数字高程模型 DEM (digital elevation model)
数字化 digitization
数字化世界标准地震台网 DWWSSN (Digital World Wide Standard Seismograph Network)
数字化文件 digital file
数字式测高仪 digisonde
数字图形处理 digital graphic processing
衰减 attenuation
衰减常数 attenuation constant
衰减系数 attenuation coefficient
衰落 fade
双标准纬线投影 projection with two standard parallels
双侧断裂 bilateral faulting
双重符合 twofold coincidence
双重核共振磁力仪 double nuclear resonance magnetometer; Overhauser magnetometer
双重星系 binary system
双极磁区 bipolar magnetic region

双极扩散 ambipolar diffusion
双极群 bipolar group
双流不稳定性 two-stream instability
双流等离子体 two-stream plasma
双麦克斯韦分布 bi-Maxwellian distribution
双曲线轨道彗星 hyperbolic comet
双式扩散 bimodal diffusion
双向通量 bidirectional flux
双向性 bidirectionality
双向异性 bidirectional anisotropy
双原子分子 diatomic molecule
双原子离子 diatomic ion
水槽 flume
水地球化学测量 geochemical drainage survey
水电离限 water ionization limit
水电站 hydropower station
水氡 water radon
水动力的 hydrodynamic
水合物 hydrate
水库 reservoir
水库触发地震 reservoir-triggered seismicity
水库地震 reservoir-induced earthquake
水力的 hydraulic
水力学 hydraulics
水面蒸发 water surface evaporation
水平分量 horizontal component
水平回线法 HLEM (horizontal loop method)
水平井 horizontal well
水平控制网 horizontal control network
水平扩散 horizontal diffusion
水平强度 horizontal intensity
水平折光差 horizontal refraction error
水圈地球化学 geochemistry of the hydrosphere
水热爆炸 hydrothermal explosion
水热对流系统 hydrothermal convection system
水热活动 hydrothermal activity
水热矿化 hydrothermal mineralization
水热喷发 hydrothermal eruption
水热区 hydrothermal area
水热蚀变 hydrothermal alteration
水热田 hydrothermal field
水热系统 hydrothermal system
水热循环 hydrothermal circulation
水热资源 hydrothermal resource
水头 water head
水土 water and soil
水土保持 soil conservation
水位 water level
水文的 hydrological
水文过程 hydrological process
水文模型 hydrological model
水文学 hydrology
水文预报 hydrological forecasting
水文站 hydrologic station
水吸收光谱 water absorption spectrum
水系 water system
水星 Mercury
水压致裂 hydrofracturing
水准 leveling
水准测量 leveling measurement
水准点 benchmark
水准路线 leveling line
水准面 level surface
水准椭球 normal level ellipsoid
顺时针回旋 gyrate clockwise
顺序量表 ordinal scaling
瞬变 transient
瞬变场法 transient field method
瞬间地图 twinkling map
瞬时变化 transient change
瞬时磁天顶 instantaneous magnetic zenith

瞬时等离子体事件 transient plasma event
瞬时极 instantaneous pole
瞬时经度 instantaneous longitude
瞬时事件 transient event
瞬时纬度 instantaneous latitude
丝网印刷 silk-screen printing
斯蒂芬定律 Stefan's law
斯特默长度 Stormer length
斯特默锥 Stormer cone
斯通莱波 Stoneley wave
斯韦劳事件 Thvera event
锶云 strontium cloud
撕膜片 peel-coat film
四极子 quadrupole
四色印刷 four colour printing
松弛源 relaxation source
松山期 Matuyama epoch
速度 velocity
速度场 velocity field
速度反应谱 velocity response spectrum
速度结构 velocity structure
速度滤波 velocity filtering
速度弥散度 velocity dispersion
速度势 velocity potential
速度梯度 velocity gradient
速率方程 rate equation
速率系数 rate coefficient
速移极光 rapidly moving aurora
塑料闪烁体 plastic scintillator
酸性地球化学障 acid geochemical barrier
算法 algorithm
随机分布 random distribution
随机加速 stochastic acceleration
随机误差 random error
随机噪声 random noise
碎屑磁颗粒 detrital magnetic particle
碎屑剩磁 detrital remanence; DRM (detrital remanent magnetization)
隧道波 tunneling wave
隧道效应〖地震波的〗 tunneling effect (of seismic waves)
损失锥 loss cone
缩微地图 microfilm map

台湾强地动一号台阵 SMART 1 (Strong-Motion Array in Taiwan Number 1)
台阵 array
太空飞行体磁场 spacecraft magnetic field
太空摄影测量 space photogrammetry
太阳 X 射线 solar X-ray
太阳参数 solar parameter
太阳常数 solar constant
太阳潮 solar tide
太阳潮汐位能 solar tide potential
太阳赤道面 solar equatorial plane
太阳磁层坐标 solar magnetospheric coordinate
太阳磁象仪 solar magnetograph
太阳等离子体 solar plasma
太阳电磁辐射 solar electromagnetic radiation
太阳电离辐射 solar ionizing radiation
太阳电子事件 solar electron event
太阳发射谱 solar emission spectrum
太阳风 solar wind
太阳风磁场 solar wind magnetic field
太阳风粒子 solar wind particle
太阳风流 solar wind flow
太阳辐射 solar radiation
太阳辐射通量 solar radiation flux
太阳光谱 solar spectrum

太阳光谱观测镜 spectrohelioscope
太阳光谱仪 solar spectrograph
太阳光球 solar photosphere
太阳光子通量 solar photon flux
太阳黄道坐标 solar ecliptic coordinate
太阳活动 solar activity
太阳活动脉冲 solar activity impulse
太阳活动事件 solar activity event
太阳活动效应 solar activity effect
太阳活动周涨落 solar cycle fluctuation
太阳活动紫外指数 index of solar ultraviolet activity
太阳活动最小值 solar activity minimum
太阳极大年 solar maximum year
太阳极大值 solar maximum
太阳控制 solar control
太阳粒子事件 solar particle event
太阳罗盘 sun compass
太阳漫辐射 diffuse solar radiation
太阳能电池 solar battery
太阳能谱 solar energy spectrum
太阳年 solar year
太阳气象关系 solar-meteorological relationship
太阳扰动 solar disturbance
太阳日变化 solar daily variation
太阳射电波 solar radio wave
太阳射电发射 solar radio emission

太阳射电指数 solar radio index
太阳事件 solar event
太阳-天气场 solar-weather field
太阳天文学家 solar astronomer
太阳同步 sun-synchronous
太阳望远镜 helioscope
太阳微粒发射 solar corpuscular emission
太阳物理学家 solar physicist
太阳系 solar system
太阳向点 solar apex
太阳相角 solar phase angle
太阳耀斑 solar flare
太阳引力场 solar gravitational field
太阳宇宙线 solar cosmic ray
太阳指数 solar index
太阳质子 solar proton
太阳质子监视仪 sun proton monitor
太阳质子事件 solar proton event
太阳中微子单位 solar neutrino unit
太阳紫外辐射 solar ultraviolet radiation
太阳总辐射 sun's total radiation
太阴潮 lunar tide
态势地图 posture map
泰罗斯卫星 Tiros satellite
坍塌检验 slump test
滩状地震相 bank seismic facies
弹性波 elastic wave
弹性回跳 elastic rebound
弹性碰撞 elastic collision
探测 exploration
探测气球 balloon-sonde
探地雷达 ground penetrating radar
探空火箭 sounding rocket
探空气球 sounding balloon
探空系统 sounding system
碳氢化合物 hydrocarbon
碳酸气孔 mofette
碳酸泉 carbonated spring
碳酸盐 carbonate

碳酸盐岩类储集层 carbonate reservoir
碳循环 carbon cycle
汤姆森-哈斯克尔矩阵法 Thomson-Haskell matrix methord
汤姆森散射 Thomson scattering
逃逸高度 level of escape
逃逸速度 velocity of escape
逃逸通量 escape flux
逃逸锥 loss cone
套管 casing
套芯钻 overcoring
特性曲线 characteristic curve
特征波 characteristic wave
特征码 feature code
特征码清单 feature code menu
特种地图 particular map
梯度风 gradient wind
梯度漂移电流 gradient drift current
体波震级 body wave magnitude
体发射 volume emission
体积电荷 volume charge
体积元 volume element
天电定位 sferics fix
天电观测 sferics observation
天电接收器 sferics receiver
天电学 sferics
天顶 zenith
天顶光弧 zenithal arc
天顶角 zenith angle
天顶距 zenith distance
天顶投影 zenithal projection
天顶仪 zenith telescope
天空背景 sky background
天气观测 synoptic observation
天气图 synoptic chart
天球经度 celestial longitude
天球纬度 celestial latitude
天然剩磁 natural remanence; NRM (natural remanent magnetization)
天体背景 celestial background
天体弹道学 astroballistics

天体动力学 astrodynamics
天体光谱学 astronomical spectroscopy; astrospectroscopy
天体摄谱仪 astronomical spectrograph
天体位置 celestial position
天体物理学 astrophysics
天体演化学 cosmogony
天体坐标 celestial coordinate
天王星 Uranus
天卫八 Bianca
天卫二 Umbriel
天卫六 Cressida
天卫六 Cordelia
天卫七 Ophelia
天卫三 Titania
天卫十 Desdemona
天卫十二 Portia
天卫十七 Sycorax
天卫十三 Rosalind
天卫十四 Belinda
天卫十五 Puck
天卫十一 Juliet
天卫四 Oberon
天卫五 Miranda
天卫一 Ariel
天文常数 astronomical constant
天文大地垂线偏差 astro-geodetic deflection of the vertical
天文大地网 astro-geodetic network
天文单位 astronomical unit
天文导航 astronavigation
天文的 astronomical
天文点 astronomical point
天文经度 astronomical longitude
天文年历 almanac
天文闪烁 astronomical scintillation
天文时 astronomical time
天文台 observatory
天文纬度 astronomical latitude
天文折射 astronomical refraction
天文子午圈 astronomical meridian
天文子午线 astronomical meridian
天文坐标 astronomical coordinate
天线 antenna
天线孔径 antenna aperture
天线匹配 antenna matching
天线阻抗 antenna impedance
填充地图 outline map (for filling)
调和分析 harmonic analysis
调角 angle modulation
调节 adjustment
跳距 skip distance
烃 hydrocarbon
烃基 hydroxyl
烃类检测 hydrocarbon indicator (HCI)
停止相 stopping phase
通道 channel
通道倍增器 channel multiplier
通道波 channel wave
通风 aeration
通信卫星 communication satellite
通用横墨卡托投影 UTM (Universal Transverse Mercator projection)
通用极球面投影 UPS (Universal Polar Stereographic projection)
同步变化 synchronous changes
同步辐射 synchrotron radiation
同步高度 geosynchronous altitude; synchronous altitude
同步观测 simultaneous observation
同步轨道 synchronous orbit
同步卫星 geostationary satellite
同步旋转 synchronous rotation
同生地球化学异常 syngenetic geochemical anomaly
同态反褶积 homomorphic deconvolution
同位素 isotope
同位素测井 radioisotope logging

同位素地球化学 isotope geochemistry
同位素地球温度计 isotopic geothermometer
同位素示踪物 isotope tracer
同震的 co-seismic
统计地图 statistic map
统计力学 statistical mechanics
统计误差 statistical error
统计涨落 statistical fluctuation
统一震级 unified magnitude
投影 projection
投影变换 projection transformation
投影变形 distortion of projection
透镜状地震相 lens seismic facies
透明等离子体 transparent plasma
透明注记 stick-up lettering
透射函数 transmission function
透射矩阵 transmission matrix
透射系数 transmission coefficient
透视截面法 perspective traces
透视投影 perspective projection
突发电离层骚扰 SID (sudden ionospheric disturbance)
突发相 breakout phase
突发相位异常 SPA (sudden phase anomaly)
图幅 mapsheet
图幅编号 sheet designation; sheet number
图幅接边 edge matching
图幅接合表 index diagram; sheet index
图廓 edge of the format; map border
图历簿 mapping recorded file
图例 legend
图面配置 map layout
图像 image
图像分辨率 image resolution
图像复原 image restoration
图像匹配 image matching
图像数据 image data
图像数据库 image database
图像镶嵌 image mosaic
图像质量 image quality
图形-背景辨别 figure-ground discrimination
图形符号 graphic symbol
图形记号 graphic sign
图形权倒数 weight reciprocal of figure
图形元素 graphic element
土层 soil layer
土地覆被 land cover
土地利用 land use
土地沙漠化 land desertification
土地资源 land resource
土拉姆法 Turam method
土壤表层 soil surface
土壤地球化学 geochemistry of the soil
土壤地球化学测量 geochemical soil survey
土壤肥力 soil fertility
土壤环境 soil environment
土壤类型 soil type
土壤理化性质 soil physical and chemical property
土壤剖面 soil section
土壤侵蚀 soil erosion
土壤微生物 soil microorganism
土壤系统分类 soil system classification
土壤性质 soil property
土壤盐分 soil salinity
土壤盐渍度 soil salinity
土壤盐渍化 soil salinization
土壤养分 soil nutrient
土壤样品 soil sample
土壤质地 soil texture
土壤资源 soil resource
土卫 Saturnian satellite
土卫八 Iapetus
土卫二 Enceladus
土卫九 Phoebe
土卫六 Titan
土卫七 Hyperion

土卫三 Tethys
土卫十 Janus
土卫十二 Dione B
土卫十六 Prometheus
土卫十三 Telesto
土卫十四 Calypso
土卫十五 Atlas
土卫十一 Epimetheus
土卫四 Dione
土卫五 Rhea
土卫一 Mimas
土星 Saturn
土星光环 Saturn's ring
土样 soil sample
湍流 turbulence
湍流边界层 turbulent boundary layer
湍流层 turbosphere
湍流层顶 turbopause
湍流磁场 turbulent magnetic field
湍流等离子体 turbulent plasma
湍流分量 turbulence component
湍流耗散 turbulent dissipation
湍流混合 turbulent mixing
湍流交换 turbulent exchange
湍流扩散 turbulent diffusion
湍流谱 spectrum of turbulence; turbulence spectrum
湍流输送 turbulent transfer
湍流运动 turbulent motion
托架 towed boom
陀螺磁罗经的 gyromagnetic
陀螺地平仪 gyro horizon
椭率改正 ellipticity correction
椭球 ellipsoid
椭球扁率 flattening of ellipsoid
椭球长半径 major radius of ellipsoid
椭球面大地测量学 ellipsoidal geodesy
椭球偏心率 eccentricity of ellipsoid
拓扑的 topological
拓扑地图 topological map
拓扑关系 topological relation
拓扑检索 topological retrieval

外层空间 outer space
外磁层 outer magnetosphere
外磁场 external magnetic field
外电场 external electric field
外电离层 outer ionosphere
外辐射带 outer radiation belt
外核 outer-core
外空生物学 exobiology
外流通量 outward flux
外推 extrapolation
外行星 superior planet
外逸层 exosphere
外源场 external field
弯道 bend
弯曲法 bending method
弯线地震 crooked line seismic
完全电离等离子体 fully ionized plasma
晚冰期 late glacial time
网点 dot; stipple
网格 grid
网格单元 grid bin; grid cell
网格地图 grid map
网格法 grid method
网格结构 grid structure
网平差 network adjustment
网屏 screen
网纹片 transparent foil
网线 ruling
威尔莫地震仪 Willmore seismograph
微巴 barye
微波辐射 microwave radiation
微电极测井 micrologging; microresistivity logging
微电阻率测井 microresistivity logging
微分 differential
微分密度谱 differential density spectrum
微分能谱 differential energy spectrum
微观宇宙 microcosm
微黑子 microspot
微粒辐射 corpuscular radiation
微粒食 corpuscular eclipse
微粒效应 corpuscular effect
微粒宇宙线 corpuscular cosmic ray
微量成分 trace constituent
微脉动 micropulsation
微震 microearthquake; microseism
微震监测 microseismic monitoring
微震仪 microvibrograph
微重力测量学 microgravimetry
韦宁迈内兹均衡 Vening Meinesz isostasy
围压 confining pressure
维恩定律 Wien's law
维歇特地震仪 Wiechert seismograph
伪捕捉区 pseudo-trapping region
伪等值线地图 pseudo-isoline map
伪加速度反应谱 pseudo-acceleration response spectrum
伪速度反应谱 pseudo-velocity response spectrum
尾波 coda
尾刺系统 tail stingers system
纬度 latitude

纬度采用值 adopted latitude
纬度差 difference of latitude
纬度-地方时分布 latitude-local time distribution
纬度校正 latitude correction
纬度效应 latitude effect
纬向波数 zonal wave number
纬向动能 zonal kinetic energy
纬向风 zonal wind
纬向环流 zonal circulation
卫星测高 satellite altimetry
卫星大地测量学 satellite geodesy
卫星轨道 satellite orbit
卫星红外资料 satellite infrared data
卫星气象学 satellite meteorology
卫星闪烁 satellite scintillation
卫星食 eclipse of satellite
卫星寿命 lifetime of a satellite
卫星寿命预报 satellite lifetime prediction
卫星系统 satellite system
卫星星下点 sub-satellite point
卫星直接探测 direct satellite probing
卫星姿态 attitude of a satellite
卫影凌行星 shadow transit
未扰动大气 unperturbed atmosphere
未扰动模式 unperturbed model
未扰动太阳状况 undisturbed solar condition
未知层 ignorosphere
位场延拓 continuation of potential field
位形图 topographic diagram
位移反应谱 displacement response spectrum
温度变化 temperature change
温度测井 temperature logging
温度剖面 temperature profile
温度效应 temperature effect
温纳排列 Wenner array
温室效应 greenhouse effect

文化地图 cultural map
紊流 turbulence
稳定捕捉 stable trapping
稳定日珥型耀斑 stationary prominence flare
稳定同位素地球化学 stable isotope geochemistry
稳定振动 stable oscillation
稳健估计 robust estimation
稳态理论 steady-state theory
稳态蠕变 steady-state creep
稳压极光红弧 stable auroral red arc
涡度 vorticity
涡度场 vorticity field
涡度传输 vorticity transfer
涡度方程 vorticity equation
涡度输送理论 vorticity transport theory
涡量 vorticity
涡旋能量 eddy energy
沃尔夫黑子数 Wolf number
沃尔夫数 Wolf number
污染物生物地球化学循环 biogeochemical cycle of pollutants
无窗光电倍增管 open photomultiplier
无磁场空间 field-free space
无磁等离子体 unmagnetized plasma
无定向磁强计 astatic magnetometer
无机的 inorganic
无机地球化学 inorganic geochemistry
无机闪烁器 inorganic scintillator
无机闪烁体 inorganic scintillator
无倾线 aclinic line
无铁陨石 asiderite
无线电相位法 radio-phase method
无线电遥测地震采集 radio telemetry seismic data acquisition

无线电遥测地震数据采集 radio telemetry seismic data acquisition
无旋波 irrotational wave
无震带 aseismic belt
无震海岭 aseismic ridge
无震滑动 aseismic slip
无震区 aseismic zone
无滞剩磁 ARM (anhysteretic remanent magnetization)
无重量 antigravity

伍德-安德森地震仪 Wood-Anderson seismograph
物理大地测量学 physical geodesy
物理地球化学 physical geochemistry
物理机制 physical mechanism
物探 geophysical exploration; geophysical prospecting
误差 error
误差来源 error source

西杜杰尔事件 Sidutjall event
吸收带 absorption band
吸收光谱 absorption spectrum
吸收介质 absorbing medium
吸收系数 absorption coefficient
吸收指数 absorption index
稀薄等离子体 rarefied plasma; tenuous plasma
稀薄等离子体 dilute plasma
稀疏区 rarefaction region
熄火高度 burnout altitude
席状地震相 sheet seismic facies
席状披盖地震相 sheet drape seismic facies
系列地图 series maps
系统误差 systematic error
下地壳 lower crust
下地幔 lower mantle
下射天波 down coming sky wave
下投式探空仪 dropsonde
夏旱 summer drought
夏至(点) summer solstice
显著水平 significance level
现势地图 up-to-date map
现用标准大气 currently-used model atmosphere
限差 tolerance
限航区 restricted area
线偏振 linear polarization
线形网 linear triangulation network
线性调频脉冲 chirp
线阵遥感器 linear array sensor
线状符号 line symbol
陷落地震 collapse earthquake
乡村规划测量 rural planning survey
相对臭氧浓度 relative ozone concentration
相对定位 relative positioning
相对定向 relative orientation
相对定向元素 element of relative orientation
相对丰度 relative abundance
相对航高 relative flying height
相对论 theory of relativity
相对论改正 relativistic correction
相对论性天体事件 relativistic event
相对论性天体物理学 relativistic astrophysics
相对论质量 relativistic mass
相对误差 relative error
相对振幅保持 relative amplitude preserve
相对重力测量 relative gravity measurement
相反极性 opposite polarity
相干 coherence
相干叠加 coherence stack
相干加强 coherence emphasis
相干声呐测深系统 interferometric seabed inspection sonar
相干载波 coherent carrier
相关观测平差 adjustment of correlated observation
相关平差 adjustment of correlated observation
相关器 correlator
相关涨落 correlated fluctuation
相似理论 theory of similarity

相速度 phase velocity
镶嵌索引图 index mosaic
向甫鲁条件 Scheimpflug condition; Czapski condition
向外通量 outward flux
向阳侧 solar side
向阳彗尾 sunward tail
向阳极性 toward polarity
向阳扇区 toward sector
向源初动 kataseismic onset
向源震 kataseism
巷 lane
巷道验收测量 footage measurement of workings
巷宽 lane width
相 phase
相差 phase difference
相位比较 phase comparison
相位传递函数 PTF (phase transfer function)
相位多值性 phase ambiguity
相位激发极化法 phase induced polarization method
相位模糊度解算 phase ambiguity resolution
相位漂移 phase drift
相位闪烁 phase scintillation
相位稳定性 phase stability
相位周 lane; phase cycle
相位周值 lane width; phase cycle value
象限仪 quadrant
象形符号 replicative symbol
像差 aberration
像差常数 aberration constant
像场角 angular field of view
像等角点 isocentre of photograph
像底点 photo nadir point
像地平线 horizon trace; image horizon
像幅 picture format
像空间坐标系 image space coordinate system
像偶极子 image dipole
像片比例尺 photo scale
像片地质解译 geological interpretation of photograph
像片地质判读 geological interpretation of photograph
像片方位角 azimuth of photograph
像片方位元素 photo orientation element
像片基线 photo base
像片纠正 photo rectification
像片内方位元素 element of interior orientation
像片判读 photo interpretation
像片平面图 photoplan
像片倾角 tilt angle of photograph
像片外方位元素 element of exterior orientation
像片镶嵌 photo mosaic
像片旋角 swing angle
像片主距 principal distance of photo
像平面坐标系 photo coordinate system
像素 pixel
像移补偿 IMC (image motion compensation)
像主点 principal point of photograph
像主纵线 principal line
消光系数 extinction coefficient
消减 mute; subduction
消减带 subduction belt; subduction zone
消减型地热带 subduction-type geothermal belt
消减噪声 mute
消散波 evanescent wave
消失线 vanishing line
消振 shock absorption
硝酸 nitric acid
销钉定位法 stud registration
小波 wavelet
小波变换 wavelet transform
小波分析 wavelet analysis
小潮升 neap rise

小角度法 minor angle method
小流域 small watershed
小区划 microregionalization; microzonation
小像幅航空摄影 SFAP (small format aerial photography)
小行星 asteroid
小行星带 asteroid belt; asteroid zone
楔状地震相 wedge seismic facies
协方差 covariance
协方差函数 covariance function
协调世界时 UTC (coordinate universal time)
协调世界时时号 time signal in UTC
挟沙的 sediment-carrying
斜激波 oblique shock
斜截面法 oblique traces
斜压的 baroclinic
斜压条件 baroclinic condition
斜轴投影 oblique projection
谐波分析 harmonic analysis
谐波简正振型 overtone normal mode
泄洪 flood discharge
泄洪道 spillway
泄漏振型 leaking mode
心象地图 mental map
新生代 Cenozoic
信标 beacon
信标卫星 beacon satellite
信标延迟 beacon delay
信号杆 signal pole
信息数据库 information database
信息属性 information attribute
信息提取 information extraction
信噪比 signal-to-noise
星际风 interstellar wind
星际航行学 astronautics
星际介质 interstellar medium
星际空间 interstellar space
星际谱线 interstellar line
星际太阳等离子体 interplanetary solar plasma
星历 ephemeris
星系际气体 intergalactic gas
星鱼辐射带 Starfish radiation belt
星载遥感器 satellite-borne sensor
行波 travelling wave
行差 run error
行星表面学 planetography
行星波 planetary wave
行星测量学 planetary geodesy
行星大地测量学 planetary geodesy
行星动态 planetary configuration
行星光行差 planetary aberration
行星际背景 interplanetary background
行星际尘埃 interplanetary dust
行星际磁场 IMF (interplanetary magnetic field)
行星际磁力线 interplanetary magnetic field line
行星际的 interplanetary
行星际等离子体 interplanetary plasma
行星际电流 interplanetary current
行星际激波 interplanetary shock
行星际间断 interplanetary discontinuity
行星际空间 interplanetary space
行星际扰动 interplanetary disturbance
行星际闪烁 interplanetary scintillation
行星际事件 interplanetary event
行星际物质 interplanetary matter
行星距离定律 law of planetary distance
行星流星雨 planetary stream
行星热收支 planetary heat budget
行星岁差 planetary precession
行星物理学 planetary physics
行星演化学 planetary cosmogony

行星震学　planetary seismology
行政区划图　administrative map
形变　deformation
形变监测　deformation monitoring
形成机制　formation mechanism
修版　retouching
修正地磁坐标　corrected geomagnetic coordinate
修正偶极坐标　corrected dipole coordinate
虚地磁极　VGP (virtual geomagnetic pole)
虚反射　ghost reflection
虚反射高度　virtual reflection height
虚高　virtual height
虚拟地景　virtual landscape
虚拟地图　virtual map
虚实分量法　imaginary-real component method
虚位移　virtual displacement
序贯平差　sequential adjustment
续至波　secondary wave
悬浮颗粒　aerosol
悬式经纬仪　hanging theodolite
旋臂　spiral arm
旋涡结构　spiral structure
旋转波　rotational wave
旋转参数　rotation parameter
旋转磁强计　spinner magnetometer
旋转角　rotation angle
旋转剩磁　rotational remanence; RRM (rotational remanent magnetization)
选取限额　nom for selection
选取指标　index for selection
选权迭代法　iteration method with variable weights
选择 $\gamma-\gamma$ 测井　selective $\gamma\text{-}\gamma$ logging
选择经度　assumed longitude
选择纬度　assumed latitude
选择效应　selection effect
熏烟纸记录图　smoked paper record
寻常波　ordinary wave
巡天观测　sky patrol
循环磁化　cyclic magnetization
循环磁化强度　cyclic magnetization
循环周期　recurrent period
汛期　flood season

压磁效应 piezo-magnetic effect
压力标高 pressure scale height
压力梯度 pressure gradient
压力验潮仪 pressure gauge
压力轴 P-axis (pressure axis)
压剩磁 piezo-remanence; PRM (piezo-remanent magnetization)
压缩波 compressional wave
压制 suppression
亚暴电流 substorm current
亚暴活动 substorm activity
亚暴强度 substorm intensity
亚极光带 subauroral zone
亚视阈极光 sub-visual aurora
亚太区域地理信息系统基础设施常设委员会 PCGIAP (Permanent Committee on GIS Infrastructure for Asia and the Pacific)
氩 argon
烟迹 smoke trails
湮没 annihilation
延伸距离 extended distance
延时组合 beam steering
延拓 continuation
岩爆 rock burst
岩浆房 magmatic chamber; magmatic pocket
岩浆环流 magmatic circulation
岩浆水 magmatic water
岩漠 hammada
岩石层 lithosphere
岩石磁性 rock magnetism
岩石圈 lithosphere
岩石圈地球化学 geochemistry of lithosphere
岩石学 lithology
岩心 core
岩性 lithology
沿岸测量 coastwise survey
沿海测量 coastwise survey
盐度 salinity
盐分 salinity
盐碱 saline-alkaline
盐渍 saline
盐渍化 salinization
颜色空间 colour space
衍射 diffraction
掩 occultation
掩日测量 solar occultation measurement
掩星 occultation of star
验潮 tidal observation
验潮仪 tide-meter
验潮站 tidal station
验潮站零点 zero point of the tidal
验震器 seismoscope
阳电子 positron
阳极 anode
阳离子 positive ion
阳像 positive image
洋底喷气孔 submarine fumarole
洋底热泉 submarine hot spring
洋脊型地震 ridge-type earthquake
洋中脊 mid-ocean ridge
仰角 angle of elevation
氧地球化学障 oxygen geochemical barrier
氧化机制 oxidation mechanism
遥测 telemetry
遥测地震台网 telemetered seismic network

遥测地震仪 telemetric seismic instrument
遥测系统 telemetry system
遥感 remote sensing
遥感测深 remote sensing sounding
遥感模式识别 pattern recognition of remote sensing
遥感平台 remote sensing platform
遥感数据获取 remote sensing data acquisition
遥感制图 remote sensing mapping
耀斑爆发 outbreak of flares
耀斑辐射 solar flare cosmic radiation
耀斑钩扰 solar flare crochet
耀斑活动 solar flare activity
耀斑激浪 flare surge
耀斑粒子 flare particle
耀斑扰动 solar flare disturbance
耀斑效应 solar flare effect
耀斑宇宙线 solar flare cosmic ray
耀斑质子事件 solar flare proton event
野外 field
野外地质图 field geological map
野外填图 field mapping
曳力效应 drag effect
曳引效应 drag effect
夜侧磁层 nightside magnetosphere
夜光云 noctilucent cloud
夜间臭氧变化 nocturnal ozone variation
夜间辐射 nocturnal radiation
夜气辉光谱 night air-glow spectrum
夜气辉连续谱 nightglow continuum
夜天辐射 night-sky radiation
液核 liquid core
液化 liquefaction
液氧 loxygen

一般共轭性 general conjugacy
一跳传播 one-hop propagation
一维模式 one-dimensional model
一氧化氮 nitric oxide
一氧化氮电离限 nitric oxide ionization limit
一氧化氮生成 nitric oxide production
一氧化碳反应 carbon monoxide reaction
伊勒瓦拉反向 Illawarra reversal
移动台 mobile station
异常 anomaly
异常地球化学梯度 anomalous geochemical gradient
异常电离 anomalous ionization
异顶差 altitude correction of zenith difference
异相 out of phase
异质性 heterogeneity
逸散层 exosphere
逸散层底 exobase
意境地图 mental map
翼梢系统 wing-tip system
因普特法 INPUT method (induced pulse transient method)
因瓦基线尺 invar baseline wire
阴极 cathode
阴像 negative image
音频磁场法 AFMAG (audio frequency magnetic field method)
银道面 galactic plane
银河宇宙线 galactic cosmic ray
银河中心 galactic center
银河坐标 galactic coordinate
引潮力 tide-generating force
引潮位 tide-generating potential
引导中心等离子体 guiding centre plasma
引航图集 pilot atlas
引力 gravitation
引力潮 gravitational tide
引力辐射 gravitational radiation

引力红移 gravitational red shift
引力位 gravitational potential
引力相互作用 gravitational interaction
引力中心 barycentre
引水锚地 pilot anchorage
引张线法 method of tension wire alignment
引震应力 earthquake-generating stress
印度大潮低潮面 Indian spring low water
印刷版 printing plate
迎风的 windward
迎风面 windward side
迎角 angle of attack
荧光地图 fluorescent map
影区 shadow zone
影响区 area coverage
影像 image
影像地质图 geological photomap
影像分辨率 image resolution
影像复原 image restoration
影像金字塔 image pyramid
影像匹配 image matching
影像融合 image fusion
影像数据 image data
影像数据库 image database
影像相关 image correlation
影像镶嵌 image mosaic
影像质量 image quality
应变积累 strain accumulation
应变阶跃 strain step
应变仪 strainmeter
应力过量 stress glut
应力迹线 stress trajectory
应力降 stress drop
应力解除 stress relief
应力位错 stress dislocation
应力仪 stressmeter
应力张量 stress tensor
应用地球化学 applied geochemistry
应用地球物理(学) applied geophysics
应用地震学 applied seismology
应用频率 applied frequency
硬辐射 hard radiation
硬架系统 rigid boom system; rigid frame system
永久磁化 permanent magnetization
涌浪滤波器 heave compensator
油藏 oil reservoir
油田开发地震 production seismic
游艇用图 yacht chart
有感地震 felt earthquake
有机地球化学 organic geochemistry
有机地球化学法 organic geochemistry method
有机闪烁体 organic scintillator
有机悬浮颗粒 organic aerosol
有限差分偏移 finite difference migration
有限性变换 finiteness transform
有限性校正 finiteness correction
有限性因子 finiteness factor
有限移动源 finite moving source
有限张角 finite opening angle
有效波 effective wave
有效地球半径 effective radius of the earth
有效地球辐射 effective terrestrial radiation
有效地球透射 effective atmospheric transmission
有效峰值加速度 EPA (effective peak acceleration)
有效峰值速度 EPV (effective peak velocity)
有效积分谱 integral effective spectrum
有效应力 effective stress
有效原子量 effective atomic weight
右手正交系 right-handed orthogonal system
诱发地震 induced earthquake
诱发地震活动性 induced seismicity

淤积　sedimentation
淤积物　sediment
余赤纬　codeclination
余辉　afterglow
余弦定律　cosine law
余震　aftershock
渔礁　fishing rock
渔堰　fishing haven
渔业用图　fishing chart
渔栅　fishing stake
宇宙 X 射线辐射　cosmic X-radiation
宇宙背景辐射　cosmic background radiation
宇宙等离子体　cosmic plasma
宇宙丰度　cosmic abundance; universal abundance
宇宙化学　cosmochemistry
宇宙空气动力学　cosmical aerodynamics
宇宙射电噪声　cosmic radio noise
宇宙线暴　cosmic-ray storm
宇宙线赤道　cosmic-ray equator
宇宙线丰度　cosmic-ray abundance
宇宙线集流　cosmic-ray jet
宇宙线膝　cosmic-ray knee
宇宙噪声吸收仪　riometer
宇宙噪音　cosmic noise
宇宙制图　cosmic mapping
雨水　meteoric water
预报　prediction
预报地图　prognostic map
预报因子　predictive indicator
预测反褶积　predictive deconvolution
预打样图　pre-press proof
预离解　predissociation
预期地球化学勘探　perspective geochemical prospecting
预制符号　preprinted symbol
预制感光版　presensitized plate
阈能　threshold energy
阈值刚度　threshold rigidity
愈合前沿　healing front

元数据　metadata
元素丰度　abundance of elements
元素迁移〖地球化学的〗　geochemical migration of elements
元素循环〖地球化学的〗　geochemical cycle of elements
原地测量　in situ measurement
原地剩磁　site remanence
原地探针　in situ probe
原地应力　in situ stress
原生磁化　primary magnetization
原生磁化强度　primary magnetization
原生地球化学异常　primary geochemical anomaly
原生气体　juvenile gas
原生剩磁　primary remanent magnetization
原生水　connate water; juvenile water
原始大气圈　primitive atmosphere
原宇宙辐射　primary cosmic radiation
原子发射　atomic emission
原子反应　atom reaction
原子火箭　atomic rocket
原子进动磁强计　atomic precession magnetometer
原子氢　atomic hydrogen
原子氢化学　atomic hydrogen chemistry
原子吸收系数　atomic absorption coefficient
原子序数　Z-number
原子氧　atomic oxygen
原子质量数　atomic mass number
原子钟　atomic clock
圆偏振的　circularly polarized
圆曲线测设　circular curve location
圆-圆定位　range-range positioning
圆-圆定位系统　range positioning system

圆柱投影　cylindrical projection
圆锥投影　conic projection
远场　far-field
远场面波　far-field surface wave
远场体波　far-field body wave
远程导航　long-range navigation
远程定位系统　long-range positioning system
远地点　apogee
远点　apoapsis
远拱点　apoapsis
远共振　remote resonance
远海测量　pelagic survey
远红外波道　far IR channel
远日点　aphelion
远谐振　remote resonance
远月点　apocynthion
远震　distant earthquake; teleseism
远震地震波　teleseismic wave
远紫外光摄谱仪　far ultraviolet spectrograph
约翰逊噪音　Johnson noise
月面测量　selenodesy
月面测量学　lunar geodesy; selenodesy
月面环壁平原　walled plain
月面环形山　lunar crater
月面学　selenography
月偏食　partial lunar eclipse
月平均海面　monthly mean sea level
月球测绘　lunar geodesy
月球轨道飞行器　lunar orbiter
月球学　selenology
月球振荡　lunar oscillation
月全食　total lunar eclipse
月食　eclipse of the moon; lunar eclipse
月下点　sublunar point
月相　phase of the moon
月震　moonquake
月震图　lunar seismogram
月震学　lunar seismology
月震仪　moon seismograph
跃迁概率　transition probability
云滴并合　coalescence of droplet
云高计　cloud altimeter
云际介质　intercloud medium
陨石触地点　earth point
陨星坑　meteorite crater
陨星烧蚀　ablation of meteorites
陨星天文学　meteoritic astronomy
陨星学　meteoritics
孕震　earthquake preparation
孕震区　seismogenic zone
运动方程分析解　analytical solution of motion equation
运动方程数值解　numerical solution of motion equation
运动线法　arrowhead method
运动性扰动　travelling disturbance
晕滃法　hachuring
晕渲法　hill shading

灾害 disaster
灾害预测 disaster prediction
再磁化 remagnetization
再磁化圆(弧) remagnetization circle
再分结构 subdivisional organization
再俘获 recapture
再入飞行器 reentry vehicle
再入轨道 reentry trajectory
再入式 reentry mode
载波 carrier
载波相位测量 carrier phase measurement
载荷潮 load tide
载荷勒夫数 load Love's number
载频 carrier frequency
暂态蠕变 transient creep
凿井施工测量 construction survey for shaft sinking
造山地热带 orogenic geothermal belt
泽德费尔德图 Zijderveld diagram
增强辐射 enhanced radiation
闸门 gate
栅格 raster
栅格绘图 raster plotting
栅格数据 raster data
窄带滤波器 narrow band filter
展开角 angle of spread
展开立体图 open cube display
站心坐标系 topocentric coordinate system
张力轴 T-axis (tension axis)
张位错 tensile dislocation
章动 nutation
章动噪音 nutation noise
章动周期 nutation period
障碍体 barrier
障碍体震源模式 barrier source model
找矿 prospecting
沼泽地震勘探 swamp seismic exploration
照相排字机 phototypesetter
照相天顶筒 photographic zenith tube
照相制版镜头 printer lens; process lens
照准点 sighting point
照准点归心 sighting centring
折光 refraction
折合摆长 reduced pendulum length
折合热流量 reduced heat flow
折合走时 reduced travel time
折射 refraction
折射波对比法 refraction correlation method
折射率 index of refraction; refractive index
折中选择研究 trade-off studies
褶皱检验 fold test
针叶林 coniferous forest
真地平线 true horizon
真反射高度 true reflection height
真高 true height
真偶极子 real dipole
真实孔径雷达 real-aperture radar
真水平线 true horizon
真误差 true error
真子午线 true meridian

振动能级 vibrational level
振动跃迁 vibrational transition
振幅 amplitude
振幅包络 amplitude envelope
振型-射线双重性 mode-ray duality
振子强度 oscillator strength
震动持续时间 duration of shaking
震害 earthquake damage
震后的 post-seismic
震级 earthquake magnitude
震级-频度关系 magnitude-frequency relation
震前的 pre-seismic
震情 seismic regime
震群 earthquake swarm; swarm
震相 phase; seismic phase
震相辨别 phase discrimination
震相识别 phase identification
震源 hypocentre; seismic source
震源参数 hypocentre parameter; seismic source parameter
震源尺度 focal dimension
震源定位 hypocentral location
震源动力学 seismic source dynamics
震源过程 focal process
震源机制 earthquake source mechanism; focal mechanism
震源机制解 focal mechanism solution
震源距 hypocentral distance
震源力 focal force
震源球 focal sphere
震源深度 earthquake depth; focal depth
震源时间函数 source time function
震源体积 focal volume
震源运动学 seismic source kinematics
震灾 earthquake hazard; seismic hazard
震中 epicentre; epifocus
震中对跖点 anticentre; anti-epicentre
震中方位角 epicentre azimuth
震中分布 epicentre distribution
震中距 epicentral distance
震中烈度 epicentre intensity
震中迁移 epicentre migration
蒸发过程 evaporation process
蒸发量 evaporation
蒸散 evapotranspiration
整流器 rectifier
整体大地测量 integrated geodesy
整体感 associative perception
整体结构 extensional organization
整体运动 bulk motion
整周模糊度 integer ambiguity
正比计数器 proportional counter
正常地震 normal earthquake
正常高 normal height
正常深度地震 normal earthquake
正常时差校正 NMO correction (normal moveout correction)
正常水准椭球 normal level ellipsoid
正常引力位 normal gravitation potential
正常重力 normal gravity
正常重力场 normal gravity field
正常重力公式 normal gravity formula
正常重力位 normal gravity potential
正电子 positron
正高 orthometric height
正规方程 normal equation
正激波 normal shock wave
正交函数 orthogonal function
正离子 positive ion
正频散 normal dispersion
正射投影 orthographic projection
正态分布 normal distribution

正温效应 positive temperature effect
正弦曲线 sine curve
正向极性 normal polarity
正像 right-reading; positive image
正形投影 conformal projection
正压 barotropy
正压流体 barotropic fluid
正压情况 barotropic condition
正压性 barotropy
正压状态 barotropic state
正演 forward
正演模拟 forward modeling
正异常 positive anomaly
正则坐标 Canonical coordinate
正轴投影 normal projection
正转 prograde
政治地图 political map
直达波 direct wave
直接电导率 direct conductivity
直接光致电离 direct photoionization
直接太阳光束 direct solar beam
直流清洗 DC cleaning (direct current cleaning)
植被 vegetation
植被恢复 vegetation restoration
植被类型 vegetation type
植物区系 flora
指南极 southseeking pole
志田数 Shida's number
制动轨道 braking orbit
制图 mapping
制图分级 cartographic hierarchy
制图简化 cartographic simplification
制图夸大 cartographic exaggeration
制图选取 cartographic selection
制图专家系统 cartographic expert system
制图综合 cartographic generalization
制印原图 final original
质底法 quality base method

质量感 qualitative perception
质量守恒 conservation of mass
质量吸收系数 mass absorption coefficient
质谱分析 mass spectrometry
质谱学 mass spectrometry
质谱仪 mass spectrograph
质心 barycentre
质子 proton
质子层 protonosphere
质子磁强计 proton magnetometer
质子反应 proton reaction
质子极光 proton aurora
质子能量流通量 proton energy flux
质子旋进磁力仪 proton-precession magnetometer
质子耀斑 proton flare
致核效应 nucleating effect
致密射电源 compact radio source
致偏磁体 deflecting magnet
滞后角 angle of lag
滞后效应 hysteresis effect
中暴 medium storm
中层大气 middle atmosphere
中磁层 middle magnetosphere
中等磁暴 moderate geomagnetic storm
中等扰动 moderate disturbance
中断 blackout; fade-out
中间层 mesosphere
中间层顶 mesopause
中间极性 intermediate polarity
中间件 middleware
中间梯度法 central gradient array method
中能粒子 intermediate-energy particle
中天 meridian passage
中微子 neutrino
中纬度 temperate latitude
中心偶极子 central dipole
中性层 neutrosphere
中性层顶 neutropause

中性面 neutral surface
中性气体 neutral air
中性色调 middle tone
中性质谱仪 neutral mass spectrometer
中央子午线 central meridian
中震 moderate earthquake
中子 neutron
中子-γ测井 neutron-γ logging
中子-超热中子测井 neutron-epithermal neutron logging
中子活化测井 neutron activation logging
中子活化法 neutron activation method
中子-热中子测井 neutron-thermal neutron logging
中子星 neutron star
中子-中子测井 neutron-neutron logging
钟差 clock correction
重带电粒子 heavy charged particle
重氮复印 diazo copying
重离子 heavy ion
重力 gravity
重力测量 gravity measurement
重力测量学 gravimetry
重力场 gravity field
重力等位面 equipotential surface of gravity
重力低 gravity low; gravity minimum
重力调查 gravity survey
重力高 gravity high; gravity maximum
重力计 gravimeter
重力加速度 gravity acceleration
重力勘探 gravity prospecting
重力梯度测量 gravity gradient survey
重力梯度带 gravity gradient zone
重力梯度仪 gravity gradiometer
重力位 gravity potential
重力仪 gravimeter
重力仪零漂改正 gravimeter drift correction
重力异常 gravity anomaly
重氢 deuterium
重水 heavy water
重元素地球化学 geochemistry of the heavy element
周期彗星 periodic comet
周日太阳加热 diurnal solar heating
轴极 axis pole
昼光效应 daylight effect
逐步接近 successive approximation
逐层模型 successive sheet model
逐点激发地震剖面法 walkaway seismic profiling
主测线 inline
主厂房抗震设计 anti-seismic design of main power buildings
主磁场 main field
主带 main zone
主导地震 master earthquake
主导地震事件 calibration seismic event
主导事件 calibration event; master event
主动段飞行 powered flight
主动源法 active source method
主动制导 active guidance
主黑子 main spot
主相 main phase
主震 main shock
助动重力仪 astatic gravimeter
助曲线 extra contour
注入边界 injection boundary
驻波 standing wave
驻点压力 stagnation pressure
驻激波锋面 standing shock front
驻留流星 stationary meteor
柱密度 column density
专题层 thematic overlap
专题地图 thematic map
专题地图集 thematic atlas
专题地图学 thematic cartography

专题制图仪 thematic mapper
专用地图 special use map
转化因子 conversion factor
转换 conversion; transformation
转换〖波的〗 conversion of waves
转换波 converted wave
转换断层 transform fault
转移轨道 transfer orbit
转移椭圆 transfer ellipse
转折点 turning point
转动参考系 rotating frame of reference
转动间断 rotational discontinuity
转动量子数 rotational quantum number
转动谱线 rotational line
追踪 tracing
锥面波 conical wave
准横传播 quasi-transverse propagation
准直仪 collimator
准纵传播 quasi-longitudinal propagation
姿控 attitude control
姿态 attitude
姿态控制 attitude control
姿态陀螺仪 attitude gyro
子波 wavelet
子波处理 wavelet processing
子午面 meridian plane
子午圈 meridian
子午圈曲率半径 radius of curvature in meridian
子夜极光 midnight aurora
子夜区段 midnight sector
紫色土 purple soil
紫外辐射 ultraviolet radiation
紫外光谱 ultraviolet spectrum
紫外离子源 ultraviolet ion source
紫外日气辉 ultraviolet dayglow
紫外天文学 ultraviolet astronomy
紫外通量 ultraviolet flux
自动标准地磁观测台 automatic standard magnetic observatory
自动多频电离层记录仪 automatic multifrequency ionospheric recorder
自动化地图制图 automatic cartography
自动绘图 automatic plotting
自动天文导航 automatic celestial navigation
自动综合 automatic generalization
自发断层破裂 spontaneous fault rupture
自发发射 spontaneous emission
自发辐射 spontaneous radiation
自发破裂 spontaneous rupture
自反向 self-reversal
自然 γ 测井 γ-ray logging
自然地图 physical map
自然电位测井 SP logging (self-potential logging)
自然电位法 self-potential method
自适应叠加 adaptive stack
自相关 autocorrelation
自相关系数 autocorrelation coefficient
自协方差 auto-covariance
自旋轨道 spinning orbit
自旋率 spin rate
自旋稳定 spin stabilization
自旋轴 spin axis
自由大气 free atmosphere
自由电子 free electron
自由分子流 free molecular flow
自由空气异常 free air anomaly
自由太阳风 free solar wind
自由振荡 free oscillation
自由-自由吸收 free-free absorption
综合地图 comprehensive map
综合地图集 comprehensive atlas
综合断层面解 composite fault-plane solution
综合录井 comprehensive logging

综合物探系统 integrated geophysical system
棕钙土棕壤 brown soil
总臭氧量 total ozone amount
总磁异常强度 total intensity of magnetic anomaly
总反应速率 overall rate
总离子通量 total ion flux
总热剩磁 total thermoremanent magnetization
总岁差 general precession
总吸收 total absorption
总阻力加速度 total drag acceleration
纵波 longitudinal wave
纵向的 vertical
纵向电导 longitudinal conductance
纵向能量 longitudinal energy
走时 travel time
走时表 seismological table; travel-time table
走时曲线 travel-time curve
走向定向 strike orientation
足球振型 football mode
阻挡时间 blocking time
阻挡体积 blocking volume
阻挡温度 blocking temperature
阻挡直径 blocking diameter
阻抗 impedance
阻抗界面 impedance interface
阻抗探针 impedance probe
阻塞环流 blocking circulation
阻止截面 stopping cross section

组合导航 integrated navigation
组合地图 homeotheric map
组合检波 geophone array
组合源 source array
组件式 GIS component-based GIS
钻进 drilling
钻井-地震相剖面图 drill seismic facies section
钻井液 drilling fluid
钻孔 borehole
钻孔形变计 borehole deformation gauge; borehole strainmeter
钻孔应力计 borehole stressmeter
钻探 drilling
最大可用频率 MUF (maximum usable frequency)
最小二乘法原理 principle of least squares
最小偏差角 angle of minimum deviation
最小偏向角 angle of minimum deviation
左旋 left-handed rotation
坐标 coordinate
坐标动量空间 coordinate momentum space
坐标方位角 grid bearing
坐标系 coordinate system
坐标原点 origin of coordinates
坐标转换 coordinate transformation

非汉字开头的词条

AE 指数　AE index (auroral electrojet index)
Ap 指数　Ap index
BL 坐标　BL coordinate
B 轴　B-axis; N-axis
C 指数　C index
C9 指数　C9 index
C-C 效应　cross-coupling effect
Ci 指数　Ci index
CN 基反应　CN radical reaction
D 区　D region
D 区负离子　D region negative ion
D 区离子化学　D region ion chemistry
Delaunay 三角网　Delaunay triangulation
Dst 指数　Dst index
DW 法　DW method
E 区　E-region
F 区　F-region
F1 层　F1 layer
F1 缘　F1 ledge
F2 层　F2 layer
GPS 测量　GPS survey
GPS 定位　GPS positioning
GPS 观测　GPS observation
GPS 技术　GPS technology
GPS 接收机　GPS receiver
GPS 控制网　GPS control network
GPS 网　GPS network
GPS 卫星　GPS satellite
GPS 信号　GPS signal
JMA 表　JMA (intensity) scale; Japan Meteorological Agency (intensity) scale
K 参数　K parameter
K 精度参数　K precision parameter
K 指数　K index
Kp 指数　Kp index
M 界面　M discontinuity (Mohorovicic discontinuity); Moho; Mohorovicic discontinuity (M discontinuity)
MM 表　MM (intensity) scale; modified Mercalli (intensity) scale
MSK 表　MSK (intensity) scale; Medvedev-Sponheuer-Karnik (intensity) scale
MST 雷达　MST radar
N 线　N line
N 轴　B-axis; N-axis; null vector
P 波　primary wave
P 轴　P-axis (pressure axis)
PL 波　shear coupled PL waves
PS 版　presensitized plate
RF 表　Rossi-Forel (intensity) scale
R 波　Rayleigh wave
SOFAR 声道　SOFAR channel (sound fixing and ranging channel)
S 波　secondary wave; s-wave
T 震相　T phase
T 轴　T-axis (tension axis)
TM 影像　TM image
WKBJ 地震图　WKBJ seismogram
WKBJ 法　WKBJ method (Wentzel-Kramers-Brillouin-Jeffreys method)
WKBJ 理论地震图　WKBJ

theoretical seismogram
X 射线 X-ray
X 射线暴 X-ray burst
X 射线背景 X-ray background
X 射线辐射 X-ray radiation
X 射线光学 X-ray optics
X 射线探测器 X-ray detector
X 射线天文学 X-ray astronomy
X 射线通量 X-ray flux
X 射线望远镜 X-ray telescope
X 射线亚暴 X-ray substorm
X 射线源 X-source

20°间断 20° discontinuity

α 径迹测量 α-track etch survey
α 粒子 alpha particle
α 能量损失 alpha energy loss
α 射线 α-ray; alpha ray
β 粒子 beta particle
β 射线 β-ray; beta ray
β 衰变 beta decay
β 因子 beta factor
γ 能谱仪 γ spectrometer
γ 射线辐射 γ-ray radiation
γ-中子法 γ-neutron method
γ-γ 测井 γ-γ logging
τ 法 τ method
τ 函数 τ function

参 考 文 献

阿勒比.牛津地球科学词典.上海：上海外语教育出版社,2002.
《地球科学大辞典》编委会编.地球科学大辞典.北京：地质出版社,2005.
《地球物理勘探词典》编写组编.地球物理勘探词典.北京：科学出版社,1976.
地球物理学名词审定委员会.地球物理学名词.北京：科学出版社,1989.
地球物理学名词审定委员会.汉英地球物理学名词.北京：科学出版社.
地球物理学专业学术翻译必备词汇. http://dict.cnki.net/essential/A012_1.html.
地质辞典编纂委员会.地质辞典.北京：地质出版社,1981.
地质物产部地质辞典办公室编.地质大词典(五)(地球物理勘探、地球化学探矿).北京：地质出版社,2005.
地质学名词审定委员会,地质学名词.北京：科学出版社,1993.
都亨,李再琨主编.英汉空间物理学词汇.北京：科学出版社,1981.
傅容珊,黄建华.地球动力学.高等教育出版社,2001.
国家地质总局探矿研究所编.英汉地球物理探矿词典.北京：地质出版社,1979.
刘光文,周恩济.英汉水文学词典.科学出版社,1985.
桑隆康、赵珊茸、李方林主编.汉英英汉地球物质科学术语.中国地质大学出版社,2008.
邵广周编译.勘探地球物理专业术语汇编(中英对照). http://www.docin.com/p73645605.html.
中国空间科学学会编.空间科学词典.科学出版社,1987.
中华人民共和国国家标准《地球化学勘查术语》,GB/T14496-1993.
中华人民共和国国家标准《地质仪器术语 地球物理勘探测井仪器术语》.DZ/T0121.7-1994.
自然地理学和测绘学专业学术翻译必备词汇. http://dict.cnki.net/essential/A008.html.

外教社英汉·汉英百科词汇手册系列

1. 保险
2. 材料学
3. 财政学
4. 测绘学
5. 出版印刷
6. 地理学
7. 地球科学
8. 地质学
9. 电子、通信与自动控制技术
10. 电子商务
11. 动力与电气工程
12. 法学
13. 房地产
14. 纺织
15. 工程与技术
16. 管理学
17. 广播影视
18. 广告学
19. 国际贸易
20. 海洋与水文
21. 航空航天
22. 核科学技术
23. 化工
24. 化学
25. 环境科学
26. 机械工程
27. 计算机
28. 交通运输工程
29. 教育学
30. 经济学
31. 军事
32. 考古学
33. 会计与审计
34. 矿业工程技术
35. 力学
36. 历史学
37. 林业
38. 旅游
39. 美术、书法与摄影
40. 民族学
41. 能源科学技术
42. 农业
43. 烹饪
44. 人口学
45. 人类学
46. 人力资源管理
47. 商务
48. 社会学
49. 生理学
50. 生物学
51. 食品科学
52. 市场营销
53. 市政工程
54. 数学
55. 水产业
56. 水利工程
57. 体育
58. 天文学
59. 统计学
60. 图书馆、情报与文献学
61. 土木建筑
62. 网络与多媒体
63. 微生物与病毒学
64. 文学
65. 物理学
66. 物流与仓储
67. 戏剧、戏曲与舞蹈
68. 心理学
69. 新闻与传播
70. 信息与系统科学
71. 畜牧与兽医
72. 药学
73. 医学
74. 音乐
75. 语言学
76. 园艺学
77. 哲学
78. 政治学
79. 知识产权
80. 植物学
81. 中医中药
82. 宗教